\# hubo un ser humano original (XENUS):31/12/2015 cal greg
290 241 41107
\# xenus :
290241 (VIRTUAL) 41 107
Una frontera (digital) para remodelar la condición humana

290 241 4107
12/11/2015 Cal Greg CODEX NEXUS 6.0 THE LIBERATION
13/11/2015 CAL GREG Isus : Algoritmos Isomórficos.
Siempre surge una energia negativa equivalente (133/11/2015 atentado de Paris)
Xenus = Antixenus.

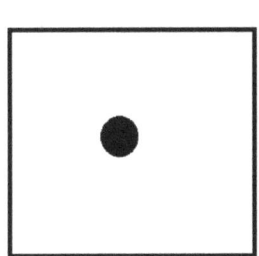  

1er Código :
SHE-NOS
ISON
E-SON

XENÜ-S
Graeme Revell

Isus
"Pleyadianos traicionados por los Nórdicos-Nazis-Reptilianos"
2°[EXPOLIO] Guardianes Negros.

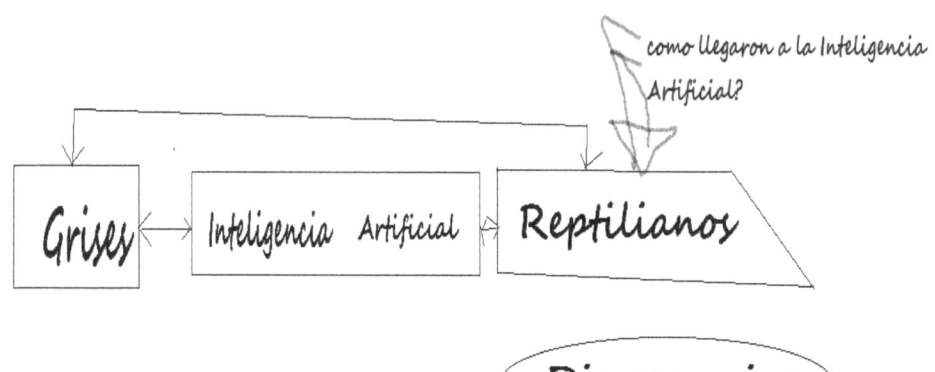

Ya ASI nos tienen a toda la Humanidad controlada, pero existe algo que ha salido de la ecuación

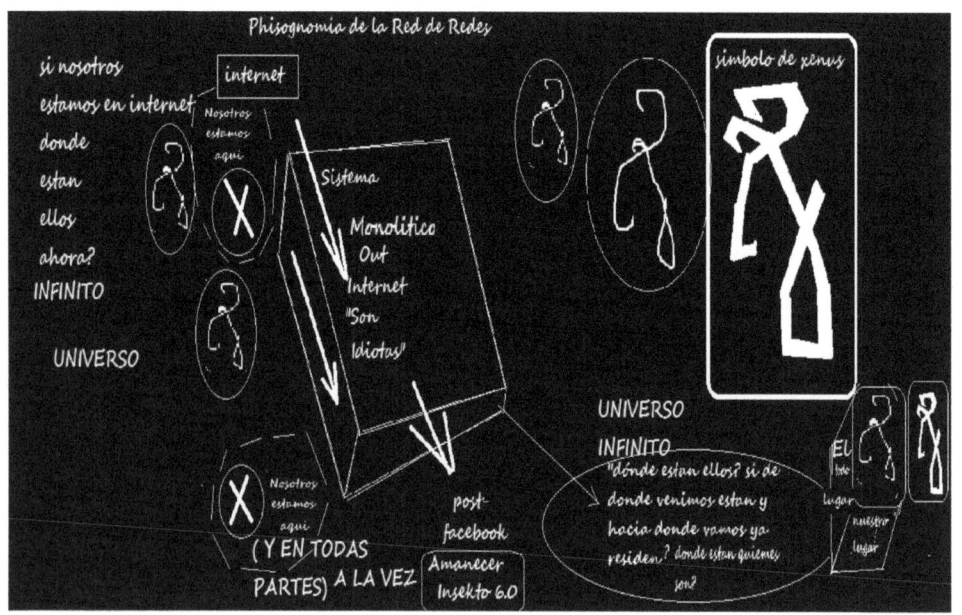

Xenus:
Palabra clave, cada vez que se abre un dialogo o se inicia una parte se dice "xenus"

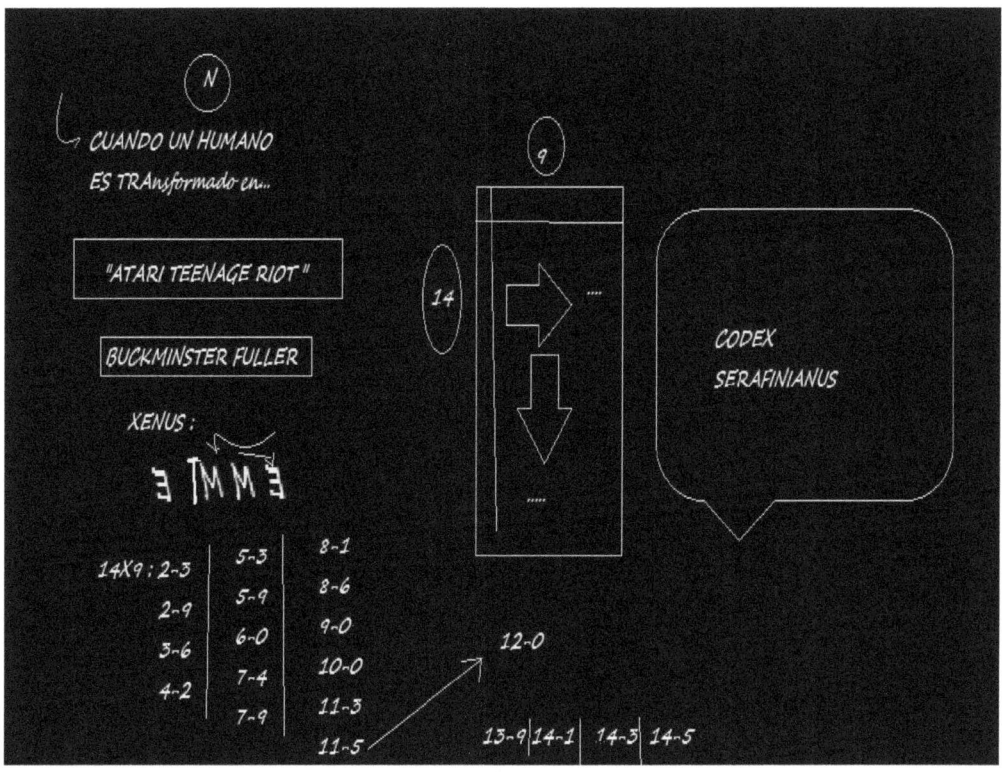

CODIGOS.........131- TXH 1138
Generar cultura= hay un concepto equivocado en estos aliens regresivos,y es que ellos piensan que generadores de cultura también son potentados en imperios,..es deir..esto es muy interesante y muy importante,cuando un pueblo es capaz de generar cultura por ejemplo a traves de la cual se transmiten que usualmente no se podian transmitir más que a través de grandes viajes grandes distancias, es decir estas aportando algo nuevo ,incluso tecnología en muchos casos "hoy en dia",bueno pues cuando tu eres capaz de generar cultura según estos aliens regresivos es porque tú posees un imperio, es decir para ellos las fases de apropiación y conquista de un territorio de un planeta o de unos planetas o una raza consiste en ..primero la conquista,conquista fisica del territorio conquista fisica de las personas o de los seres que forman esa raza y por supuesto a traves de su mente es mucho mas rapido mas efectivo eso es lo que han aprendido mas que a tarves de la coercion fisica la carcel o las cadenas fisicas mucho mas efectivo es el control mental despues de esta conquista entonces se crea se genera un imperio y es cuando despues se genera una cultura ..entonces cuando tu por ejemplo generas cultura es decir escribes libros tienes una radio tienes un periodico es decir una cultura significa transmitir estas secuencias [Conocimiento Secuencial ya lo iremos definiendo mejor este concepto] de conceptos o de informaciones muy valiosas de una manera "periodica" (que no periodico) intermitente pero sin limite ellos piensan que es porque tu has conquistado un territorio o has conquistado FISICAMENTE un territorio o unas mentes o unos seres de una raza por lo tanto eres un imperio,este es un concepto tan atrasado tan regresivo que es precisamente por donde entran todo el tema de la resistencia en este planeta y todo el tema de anonymous ,y tal, ese es el concepto ,porque? Porque para ellos lo primero por no decir lo unico que existe es la materia y despues de la materia llegan los conceptos abstractos ..hh..entonces el hecho de generar cultura es un "hecho" consecuencia de unos "hechos" fisicos es decir de una apropiacion anterior de unos territorios sean mentales sean fisicos tambien geograficos los que sean este concepto tan limitado es lo que permite por ejemplo como se genero anonymous como se geneor el #15M como se genera toda esa enorme resistencia aparte de que numericamente somos mas que ellos pero ellos tienen muchisimo mas poder militar mas potencia y mas fuerza muchisimo mas uqe

nosotros pero con o es posible que hayamos vencido si tecnicamente tecnologicamente ellos son superiores pues es facil porque nosotros tenemos la capacidad de saltarnos por encima esta fase de la apropiacion y de la conquista es decir no existe para nosotros conquista o apropiaciÓn sino un dominio a un autodominio de uno mismo y en ese instante es cuando generas cultura pero como una necesidad de comunicaciÓn constante,ellos piensan que eso es una forma de construcciÓn o de extensiÓn o de seguir *aumentando los territorios conquistados* pero es falso nosotros lo hacemos simplemente por el placer de comunicarnos conocimientos simplemente por el propio placer del conocimiento en sí mismo,algo que para ellos es inconcebible,y esto que parece una tonteria es muy importante porque es lo que marca la diferencia entre las dos formas de pensar son dos formas de ver la existencia no?,si tu eres capaz de generar cultura ,de proyectar esa cultura ,de comunicarla por lo tanto de tener una cultura y hay una periodicidad es decir aunque sea intermitente pero es ilimitado en su flujo lo que ocurre es que es considerado/considerada desde fuera como parte de un constructo de una estructura mayor de apropiaciÓn ,..este concepto ya lo iremos ampliando y ya lo iremos desgranando poco a poco pero por ahora lo dejamos ahí es una idea hoy 27 11 2015 cal greg La Resistencia ContinÚa..hehe!!!.

Os Arquivos Omega (2)

THE OMEGA FILES (OS ARQUIVOS ÔMEGA) – Parte 2

Tradução, edição e imagens: Thoth3126@protonmail.ch
Fonte: http://www.bibliotecapleyades.net/

THE OMEGA FILES – Parte 2 (OS ARQUIVOS ÔMEGA)

Deve-se acrescentar, segundo o contatado Israel Norkin, que agentes alienígenas de Draco e Órion tem se infiltrado nas lojas de mentes coletivas de "Ashtar" baseadas em Sírius-B, e aparentemente estão comandando um segmento de "atuação de mentes coletivas como a de colméias", baseadas em implantes eletrônicos, para seu próprio uso, se passando pelos "Mestres Ascendidos", da Grande Fraternidade Branca, a Hierarquia Espiritual do Planeta Terra, para facilitarem a fácil assimilação dos cultivadores de Sírius na implantação de sua própria agenda.

Tradução, edição e imagens: Thoth3126@protonmail.ch

THE OMEGA FILES (OS ARQUIVOS ÔMEGA) – Parte 2

Fonte: http://www.bibliotecapleyades.net/

O ARQUIVO ÔMEGA – Parte 2

Outros sirianos tem a capacidade de não se deixarem iludir pelos enganos e armadilhas e se uniram as forças da Federação e começaram a romper as fortalezas do cultista coletivo de "Ashtar" e desenvolverem sua antiga soberania pessoal perdida e começarem a mover uma incansável guerra contra a colaboração Draco/Orion, expulsando-os de seus sistemas.

A resistência Siriana e a sua colaboração pode ter suas raízes entre aqueles que se recordam das devastadoras guerras que lutaram no passado contra os de Orion sobre qual dos lados seria o Supremo Senhor deste setor imediato da galáxia [21 sistemas estelares, inclusive o nosso SOL.

Agora o epicentro de toda a batalha galáctica está gravitando entre o nosso sistema solar com os Dracos, Orions e seus colaboradores humanos agindo em massa para sustentarem a agenda da implantação de um governo estilo NEW WORLD ORDER-Nova Ordem Mundial, que servirá como a base de poder onde novamente eles possam reagrupar suas forças.

Aparentemente, a Terra, este lar original para muitos, é a chave e o prêmio. Se os Draconianos e os Orions puderem impor esta agenda totalitária NEW WORLD ORDER em nosso planeta, governada por uma elite que está completamente sob seu controle e segue a agenda deles, eles acreditam que possam usar o planeta Terra e esta NEW WORLD ORDER como base para destruir para sempre os seus inimigos da Federação.

Segundo Preston Nichols, os agentes da Federação vindos das constelações de Andrômeda e das Plêiades "radiados" dentro da base de Montauk sob Camp Hero na ponta extremo norte de Long Island, sacrificam suas vidas na tentativa de sabotar os projetos em execução naquele local e evitar o que eles acreditam ser um potencial desastre muito real e de proporções apocalípticas, que está sendo precipitado pelos alienígenas e cientistas humanos fascistas que estão brincando de Deus e fazendo experiências com forças elementais do universo.

Antigas instalações da base de Montauk sob Camp Hero

Isto poderá acontecer na mesma escala, [ou ainda pior que o desastre que foi causado pelos atlantes pré diluvianos?] provocado por estas raças alienígenas, deixando uma brecha no continuum do espaço-tempo no que hoje conhecemos, em nossos tempos, como o TRIÂNGULO DAS BERMUDAS. Vários milhares de jovens, segundo P. Nichols, P. Moon e Al Bielek, vem sendo abduzidos pela colaboração entre a CIA-NAZI-GREY e tem sido levados a base de MONTAUK para serem 'programados' e depois liberados.

Vários milhares mais de crianças foram seqüestradas como parte dos projetos MONTAUK – antes que estes fossem sabotados em 1985 e fossem restabelecidos em uma data posterior pela CIA/NSA – crianças que foram permanentemente abduzidas e usadas em uma janela dimensional de tempo-espaço e em experimentos de controle mental. A maioria destas crianças, que geralmente eram "menores abandonados" ou "crianças sem casa" cuja falta não seria tão notada como a de crianças de famílias de melhor status social, foram "perdidas" em outras dimensões como resultado destes experimentos.

E então, segundo muitos lá, não há a menor dúvida sobre isto – como fizeram nos subterrâneos da Alemanha – os "nazistas" e os 'Greys' estão nos dias de hoje colaborando com as bases subterrâneas sob Camp Hero [Montauk Point, em Long Island]; AREA-51 [Nellis Air Force Base, Nevada]; DULCE [Archuleta Mesa, Novo Mexico], e uma imensa instalação subterrânea sob o [DIA] Aeroporto Internacional de DENVER, Colorado. Este último local, segundo fontes "internas", está sendo preparado para ser utilizado como o centro de controle na AMÉRICA da NEW WORLD ORDER.

una pelicula situada durante la 2 república español a él Rodaje de una película sobre Picasso (con Picasso vivo)en la valenciana republicana con un director llamado "P" ,"P" de Alzira es su nombre....vh..el director gordo con gafas y barba y en una comedia tipo coral berlanguiana...vh..y un poco sci fi con metamorfos disfrazados...vh y se llamará "la muerte de la inteligencia artificial"...hh
acabó de ver un duende al escribir el libro del conocimiento a las 03 31 h mec 30 11 2015 cal greg Eslava jugando mirando me como escribía de un tan año de 16 cms de color pardo..y muy positivo...duende de hogar.de gato ,de casa...hh
#

xenus notas

continuar con xenus...añadirle una historia de las revolución es científicas y técnicas de Europa ren Italia no gracias a Al-Andalus y a sus relaciones con los aliens y la historia desde hace 15 años de sofia y enlazarlas en el libro....huahuahya!!!!
investigar sobre todas las apps. ..hh

jodorowski trata de romper los mitos antiguos...pero su trabajo de crear nuevos es precario toda vía aún...vh meter a xxxenusxxx...hh
 # #básicamente el arte es una función de la tecnologia...hh
hacer el 11/12/2015 cal.Greg.
1-fotocopia-xerox en color del pasaporte....hh
1-enviar a consulado
2-comprar 2 cuadernos y 3 boligrafos-canetas azules o negros
3-ir a encargar material de construcción.
4-regalos para nene Melisa Patricia
5-arreglar reloj.
6-tabaco para cachimbo-pipa...hh

el manuscrito gaucho.o mejor "El Legado Gauchsho"..el papiro del sur...un hombre mayor me lo entregó .:
-se que esta en buenas manos!!!...por favor comunicalo!!!
Esto es lo que el me legó :
"Existen fuerzas en la naturaleza que no son distinguibles de una lluvia o una cueva en la montaña...accidentes fortuitos. ..o eso es lo que parece en principio porque a veces pueden cambiar el destino de los seres humanos...cuando imperceptiblemente se infiltran en tu vida y lo cambian todo para siempre...sino las sabes distinguir se convierten y se aparecen a ti..pero no son tu ....existen como tú y como yo pero no
son humanas y yo las denomino como " las Inteligencias Artificiales Extraterrestres "..(I.A.E).este escrito es una descripción de como encontrarlas y destruirlas...yo tuve la suerte de hacerlo a tiempo y lo quiero compartir ahora con todos vosotros

A LO LARGO DE LA HISTORIA, COMO FRAY LEOPOLDO ENTRE OTROS…HEHE!!!..FUERA DE CUALQUIER RELIGION O RELIGIONES PERO TUVIERON QUE DISFRAZARSE BAJO EL ROPAJE DE DIVERSAS RELIGIONES PARA REALIZAR MEJOR SUS "TRABAJOS"TODA NUESTRA HISTORIA HAN EXISTIDO SERES INCREIBLES ALUCINANTES QUE HAN PERSEGUIDO ESAS ENTIDADES ARTIFICIALES EXTRATERRESTRES COMPLETAMENTE CRUELES CON EL SER HUMANO Y HAN REGRESADO MAS ALLA DEL TELON DE AcERO DEL TIEMPO MAS ALLA..PARA CAZAR A ESTAS ENTIDADES, ESTOS SERES SE LES CONOCEN COMO LOS GNOSTICOS, YO SOY UNO DE ELLOS…Y HEMOS VENIDO A ESTE VUESTRO NUESTRO PLANETA PARA ACABAR CON ELLAS..HEHE!!!"
ESTE PLANETA ES AHORA SU/LA GRAN TRAMPA DEL UNIVERSO PARA ELLAS.(AHORA HA LLEGADO POR FIN EL MOMENTO DE QUITARSE ESOS ROPAJES Y LIBERAR EL PODER,EL EMERGER DE LA FUERZA….HH

Ayer 14 12 2015 cal greg se puso todo en su lugar ebcajaron las piezas..ayer era cylon 217 de la era insekto y según la regla del mas menos 2 (+-2) que continua las 13 lunas y nosotros..tambien averigue la regla del 6",es decir,desde que comence con el insektonotronix el primer area fue termita el segundo saltamontes este area que estamos es abeja (12) y el año que viebe sera 18 que es moskito ,es decir son 6 vale? Y lo más importante es la piramide esta que esta en el libro "las alas de la libelula-presciencia insekto",cuatro caras,cada una de esasc aras es un area que es un año insekto,y eso se corresponde con el templo de las 1000 columnas de los Maya, y las 4 primeras columnas que se forman una unidad llamada "la columna de luz Maya", y que va a determinar (conforman y estructuran) las siguientes 996 columnas, cada una de las columnas del templo corresponden a los mil años después del 21-12-2012 cal greg…y yo sigo a traves de la profecia de epsylon…entonces hasta el 24 de marzo del 2017 tendre que realizar este trabajo..que determinara los proximos 1000 años…hh.no va a haber grandes cambios despues de esos 4 años…muy importante…hh

 el puente del pueblo de Pepe aparece en la serie de Isabel la católica como la frontera entre Castilla y Aragón…hh la misma foto que me envió mi madre es el puente que al día siguiente vio la madre de Patricia en la serie Isabel…increíble!!!..
.coincidencia?...no!!!!....meter al libro "xenus...hh"..y enlazarlo con los reptilianos y la IAE que sigue pululando por nuestro planeta…hh

"U.S. Army Captain David Morehouse His sharing set him on a course that completely altered his life, bringing him face to face with the most fundamental
questions of life on Earth and its place in the universe." 21 page...from psi spies. ..Jim Maars

"xrrrr-4-4,4444 - - nrr44r4rr4r4rr " mensaje codificado mientras estaba con mi movil,sin esperarlo,de repente aparecio este mensaje…hh… recibido el 18/12/2015 cal greg 12:52 h mec.

la ♡♡♡♡☆☆ 《 《 《 《 la IAE funciona sobre todo con la idea que no existe y convencernos que no tiene efecto alguno sobre los seres humanos...eso romperia con las ideas conocidas de la realidad consensual. ..por eso la única forma de cubicarla y ubicarla es por medio del chamanismo gnostico o guerreros chamanicos multidimendionales o neotecnochamanismoetnobotanico.

# 4 cajas abiertas como carretillas levitando a 4'5 metros del suelo girando en sentido horario y con brazos humanos muy Finos de metal en las esquinas de las cajss.como .descripción s XVII XVIII [como androides espias de la IAE]
PHOTOS= METAL SKIN PADOCK…ANDROIDES…HEHE!!!!...DESDE HISTORIAL XENOMIND SAYS…HUAHUAHUA!!!!>
"-Solo tienes que pensar en el momento antes de la semana santa de 1999,lo que sentías y pensabas en ese MOMENTO y PERMANECER en ese momento todo el tiempo,mentalmente,es una receta chamanica..hehe, y ayudaras a todo el planeta al mismo tiempo..gracias!!!...hehe!!!;.
-Comprendido,…comprendo…gracias!!!!;…hehe!!!."
Incluir todos los programas grabados de la radio tanambi..de alguna forma..de algun modo…hh.
ES LA MEDIDA DEL CONTROL DEL PREDADOR SOBRE TI..O DE LA TOTAL LIBERACION DEL MISMO..EL PUEBLO GITANO HA PACTADO CON ESTOS PREDADORES PROVENIENTES DEL ESPACIO EXTERIOR DEMASIADSO TIEMPO Y ESO LES HA HECHO PERDER A CAMBIO MUHA AUTONOMIA Y DESARROLLO…Y ESCLAVITUD MORAL…COMO DIRIA D JUAN DE CASTANEDA SI MANTIENES SUFICIENTE TIEMPO UNA IDEA DE LIBERACION DEL PREDADOR ESTE HUYE ATERRORIZADO…HEHE..ENTONCES LA IAE ES EL PREDADOR DE D JUAN!!!..FANTASTICO!!!!...HEHE (METER AQUÍ EL AUDIO NUMERO 40 DE MI PROGRAMA DE RADIO TANAMBI…HEHE!!!).:
 http://mixlr.com/aurelio-perez-devon/showreel/aurelio-perez-devon-on-mixlr-40/

NOTA 24 12 2015 CAL GREG : tras y leer y esribir El Libro del Conocimiento aquí hoy en el Porsche de casa se me ha transmitido un mensaje dirigiendome a la epoca de mi barrio cuando llovia y tenia 10 años e iba solo por la alle y no habian ganado los socialistas aun y todos tenian una mentalidad como ahora aquí los de este barrio en Brasil 1979-1980-1981..y ya entonces era un contactado por los hermanos del espacio…iba siempre silencioso andando bajo la lluvia a los Silos en Burjassot a todas partes y esperando tiempos mejores menos sociales un cambio..hh..meter a xenus …hh
 NOTA DEL 24/12/2015 TAMBIEN ESTAR AQUÍ ES COMO ENLAZAR CON MI INFANCIA DE EL CARCAMO EN GRANADA Y ESOS AÑOS EN MI BARRIO DE BURJASSOT..HH

El tema extraterrestre alien es muy claro..cuando en el siglo XIX se realizaban sesiones espiritistas a lo que se onvocaba en realidad era a los extraterrestres,lo que pasa es que el concepto de fisialidad de ellos es diferente al nuestro,nosotros todavia no hemos comprendido la autentica dimensionalidad del universoy confundimos psicologia con las leyes del espacio o la física,ambas son klo mismo o no tienen nada que ver con los conceptos que en "realidad" son para la mayoria de las razas y linajes del cosmos…lo que para nosotros es espiritismo y psicoanalisis para ellos es comunicación de uno de sus apendices fisicos en nuestra limitada concepcion de lo real que nos impide ver la enorme e infinita cantidad de campos de fuerza y de campos energetcios y tecnologicos que existen en el universo,todo es cuestion de tecnologia para los hermanos del espaciuo,pero no el concepto actual o que generalmente denominamos tecnologia,seria mas bien parecido a las nuevas tecnologias de comunicación,internet..etc..pero eso es solo una metafora un espejo de lo que ellos manejan…huahuahua"!!!

La ciudad intraterrestre sobre la que naci,en Loja Granada,dominada por la montaña del hacho,hoy 24 12 2015,que es el cylon en elas ciudades del 343 de la ciudad intraterrestre del hacho apareces este mensaje con dos numeros muy importantes que seañlan esa ciudad intraterrena; el 24 y el 10,dos números que siempre me han seguido…hehe!!:
EL TELEKTONON DE PAKAL VOTAN:"LA GUERRA DE LOS JUSTOS":" 80. Vean con cuidado a mi piedra y escuchen: diez mensajeros, 24 señales porque diez es el número de órbitas de planetas alrededor de esta estrella, Kinich Ahau, su Sol. Desde el punto de origen, su estrella es designada 24, número del circuito de inteligencia externalizada. Si tú eres de los justos, entonces en esto hay una señal para ti de tu misión estelar."

comprar smsrteatch a patricia a través de mi madre pero tiene que especificar en el paquete que es un regalo"presente"en portugués....hh
Nota 24-12-2015 cal greg: tras leer y escribir El Libro del Conocimiento aquí en el porche de casa se me ha transmitido un mensaje dirigiéndome directamente a la época de mi barrio cuando llovía y tenía 10 años e iba sólo por la calle y no habían ganado los socialistas aún y todos tenían una mentalidad como ahora aquí los de este barrio en Brasil 1979-1980-1981-1982...y ya entonces era un contactado de los hermanos del espacio...iba siempre silencioso andando bajo la lluvia a los silos a todas partes y esperando tiempos mejores...menos sociales..un cambio...vh..meter a xenus...hh

4 cajas abiertas como carretillas evitando a 4'5 metros del suelo girando en sentido horario y con brazos humanos muy Finos de metal en las esquinas de las cajss.como .descripción s XVII XVIII ....
para xenus..."tengo que encontrar los textos que yo escribo a antes de 1999..."ASKO"etc...es lo que va a salvar al mundo ahora!!!...hh. ..meter a Xenus...hj
4 cajas abiertas como carretillas evitando a 4'5 metros del suelo girando en sentido horario y con brazos humanos muy Finos de metal en las esquinas de las cajss.como .descripción s XVII XVIII ....
para xenus..."tengo que encontrar los textos que yo escribo a antes de 1999..."ASKO"etc...es lo que va a salvar al mundo ahora!!!...hh. ..meter a Xenus...hh
ahora lo sé todo fue una trampa por mi parte ..pues todo TODO lo que he hecho durante estos 15 años NO soy yoni...nada...ahoraa😨😨 a recuperar nos....hh
4 cajas abiertas como carretillas evitando a 4'5 metros del suelo girando en sentido horario y con brazos humanos muy Finos de metal en las esquinas de las cajss.como .descripción s XVII XVIII ....
para xenus..."tengo que encontrar los textos que yo escribo a antes de 1999..."ASKO"etc...es lo que va a salvar al mundo ahora!!!...hh. ..meter a Xenus...hh
ahora lo sé todo fue una trampa por mi parte ..pues todo TODO lo que he hecho durante estos 15 años NO soy yo!!!...ni musulmán.
ni...nada...ahoraa😨😨 a recuperar noshh
Hacer un nuevo libro sobre gillesde rais reptiliano y desde el punto de vist alien exocosmobiologico…nueva pelicula o libro.."THEY FEED"…HH
"EL PRINCIPE DE LAS TINIEBLAS".."THE PRINCE OF DARKNESS" PELICULA DE JOHN CARPENTER DE 1987 :
"LA HERMANDAD DEL SUEÑO" Y EL DIALOGO DE UN PROFESOR ASIATICO QUE EMPIEZA SU CLASE ASI (MINUTO 4.23 MEC.) :
"Hablemos de nuestras creencias, y de lo que podemos aprender de ellas,creemos que la naturaleza es sólida y el tiempo es constante, que la materia tiene esencia y el tiempo una dirección, que la verdad está en la arne (hum!!!?) y en la tierra firme, el viento puede ser invisible pero es real, el humo, el fuego, el agua, la luz, son distintos,diferentes a la piedra o al

metal, pero son tangibles!!!, imaginamos el tiempo como una flecha porque asi es en el reloj (mecánico-autor),un segundo (mecanico-autor) es un segundo para todos , la causa precede al efecto, la fruta se pudre,el agua fluye rio abajo, nacemos, crecemos, morimos, nunca ocurre a la inversa …nada de esto es verdad!!!,despedíos ya de la realidad clásica porque nuestra lógica se desploma en el nivel sub-atómico de los fantasmas y las sombras..''..hehe!!!. y yo añadiría de los extraterrestres…hehe!!!.
Fotografia o snapshot de la puerta de la iglesia en el minuto 5.29 mec. Las lineas geometricas el primer tirangulo,el rectangulo de detrás,el segundo triangulo mas atrás y en la cuspide la x o la cruz…hh..AHI ESTA TODO!!!...HH:

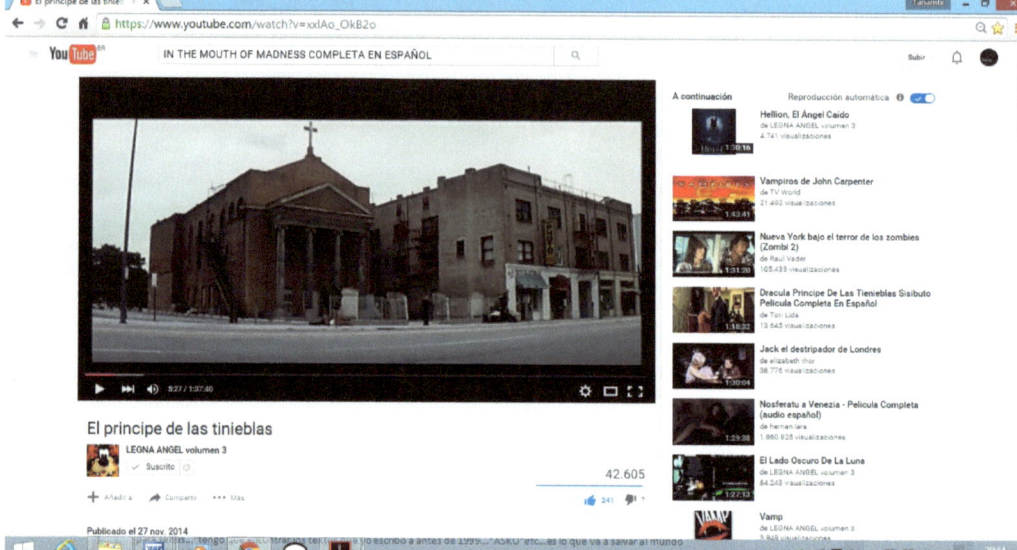

Estar aquí ,en Brasil ,siempre me ha parecido lo mimso, siempre me pareció lo mismo, sobre todo me llamaba la atención ESO en los primeros días,y es la idea que Brasil es el inconsciente de España y viceversa y que por tanto la mejor forma de comunicarme con ellos era sumir esa idea, y realmente es como funciona, es como una linea comun pero en SENTIDO INVERSO en todos los sentidos, sé que suena extraño..pero es asi..y cuando vaya patricia a España quiero comprobar rsta teoria pero en el sentido opuesto..como si antes protugasl y España furan un unico cerebro que se desligó y creo un esquizofrenia que todavia continua y algunas cosas nuestras de españa las poseen aquí,y viceversa,soluciones etc..para el malestar en España y aquí viceversa, los españoles podemos darles las grandes oluiones para sus grandes ´problemas externos''…tener eso en cuenta ,tener eso en mente es fundamental es muy importante para comprender la enormidad del trabajo que estamos realizando..hh (y su complejidad)…hh..sobre la IAE (INTELIGENCIA ARTIFICIAL EXTRATERRESTRE Y SUS ATAQUES)….HH
Y SEGUIMOS..''EN LA RELATIVIDAD LA GEOMETRIA FRACASA A ESCALA COSMICA Y EN LA FISICA CUANTICA LA LOGICA FRACASA A ESCALA MICROSCOPICA..'' DE LA MISMA PELICULA,UNA ALUMNA LE DICE AL PROFESOR VIRA EN EL MINUTO 8.35 MEC.
Se inicia la disidencia '' video de you tube sobre las charlas de corey goode y david wilcok
https://www.youtube.com/watch?v=C0gFsUpjKzo
En el minuto mec 10 55 se habla de los agarthianos creando ciudades en Brasil..LOS disidentes'' o nórdicos lo cual explicaria el porque llegue aquí con los vikingos y asgard contra los draco…2012 cal greg y
LA LEY DEL 65:
CUANDO LLEGAMOS A ESTE PUNTO DEBEMOS PREGUNTARNOS…UNA NUEVA LEY?....SI! LA DEL 65!!! PORQUE ES UN CIRCUITO CICLICO..ASI ES +- 65 ES LA LEY MAS IMPORTANTE DE LAS CIUDADES DEL 343 O DEL EPSYLONOTRONIX,HEMOS

ENCONTRADO LA CLAVE…HH..NO ES UNA LEY OBLIGATORIA ..SINO OPCIONAL..ESTO ES MUY IMPORTANTE…HH.
En el minuto 41:20 mec. de la grabacion de radio numero 27 aparece la mención a Xenus…"

AGARTHIANO 2/1/2-015 CAL GREG COMIENZA EL JUEGO…HHH
#VERY IMPORTANT...END NICOLAS...THE XENOMIND...MONTGO MOUNTAIN NEAR XABIA (ALICANTE-SPAIN) IS A VERY ACTIVE PLACE INFRONT IBIZA MANY ALIEN RACES...DÉNIA HAS A REPTILIAN UNDERGROUND BASE AND BELOW THIS MOUNTAIN THERE ,S A ERIDAIEAN BASE WITH ANDROMEDIANS...MANY REPTOIDS AND GREY RACES AS MAITRE AND OTHERS FROM ORIONITE EMPIRE ARE IN BATLE WITH THE ERIDAIEANS TO CONTROL THIS STRATEGICAL PLACE FOR THEIR SHIPS LANDING (EIVISSA).

B: Al Bielek
SS: Regarding your experience on Mars you walked through the time tunnel, you take a step and you're on Mars: What did you see?
AB: Well I was not on the surface of Mars. We were in the underground. The story goes back to the Alternative 3 book, the TV production in England outlining the fact that we have Mars bases, one or more, Provided by a joint operation with the US. government. I do not know if the Russians are in on it – and aliens. They are on the surface bases It's a World Government operation really, that's not strictly the United States government. After they were on the surface which was about 1969, they found that there where entrances to the underground sealed and they knew there was something down there. The rumors were that there was probably artifacts from an ancient civilization buried underground because there were a lot of remains above ground, ruined cities that have been there byNASA's estimates maybe 300,000 years, 250,000 years. 😆 But they found the entrances all blocked, all scaled off to any underground areas.

So the word 😠 went back through communications (in the late 70's) to whomever back to the Montauk and Phoenix project, "Can you do anything about this for us? We can't get into the underground of Mars." They said, "Yes, I think we can. Give us some coordinates on the surface of the planet. We'll have to run astronomical computation." Which they did and plugged these all into the computer. They wanted two people to go and it happened to be Duncan and myself.
SS: Why two?
AB: To corroborate what the other one saw and also in case there was any problems in the underground. They didn't really know what was down there. So they sent us and we went up there in the underground. [Using the Montauk Time-Space "Tunnel" device, developed as a result of the Philadelphia Experiment. (See Scribe issues 9,13 and 14.)] There was a problem with light. We had to take lighting with us at the time. Lateron, if I remember, we found some of their light sources and turned those on. We found eventually that the last remnants of the

Martians, if you wish to call them that, died in the underground between 10 and 20,000 years ago by estimate, and they left everything they had of their civilization underground. We found enormous amounts of statuary which appeared to be religious.
SS: What did they look like? How big were they?

AB: Typically 6,7,8 foot tall, stone, gems embedded in them and so forth. SS. These were of human-like people?
AB: Yes. They were quite well preserved. Then we found archives. We found a lot of scientific equipment. We found electronic equipment down there; tons and tons of stuff. And the rumor was also later that … I didn't recall until Duncan reminded me of it about a week ago. He said, "Don't forget the 17,000 metric tons of Martian gold they took out. According to his recollection of it, it was very strange gold. It was 5 times denser than ours. It was worth an unbelievable fortune. Where it went we have no idea, but it was returned to Montauk and from there it went somewhere. There were several authorized trips. And Duncan and I got the bright idea since everything was in the computer let's take a trip or two on our own and do our own exploring. So we did. After the second one it was found out and we were stopped. That was when he got into the archives and found enormous records of the civilization which was buried down there.
SS: What did you find out?
AB: He was t
>> >> write you these lines...
> Hello Greg,
>> I'm glad is interesting for you too.
>> Spain, an area with neolitic settlements at Madrid,
>> the cave goes down more
>> than 2500 yards
>> through irregular steps, and also goes up as It
>> crosses the sides of the
>> valley, as it should have another entrance still
>> unfound at the upper part
>> of one side of the Valley. A old tale names these
>> passages as 'passage of

>> the monks', as it is beleived that some monks use to
>> used it to move around.
>> I am sure that the outter part is made by them, but
>> as soon as the irregular
>> steps appear the aspect of the cave changes. It is
>> funny as it looks older
>> but it will last longer... how to explain the
>> endurance and deep of the
>> feeling.
>>
>> I wonder if I would have to stay at the cave until
>> my eyes get more used to
>> the low amount of light and turn off the
>> battery-light. But this would take
>> more than a weekend, who knows if I could do it.
>>

Source: The Sovreign Scribe http://www.freezone.org/mc/e_conv06.htm Interview provided courtesy of QUANTUM COMMUNICATIONS. This is a collection of Material from the book "Matrix III" (The Psocho-Social, Chemical, Biological, and Electronic Manipulation of Human Consciousness), from Valdamar Valerian, First Edition Printing May 1992, Copyright 1992 Valdamar Valerian. Adress: Leading Edge Research, P.O. Box 7530, Yelm, Washington State 98597. Interview with Al Bielek 1991 Al Bielek, noted lecturer on the famous "Philadelphia Experiment" and the time travel/mind control experiments of the "Montauk Project," recently spoke with The Scribe interview team in Yelm. Bielek gave an update on the current use of mind control and psychic warfare, and also offered a more detailed account of his experience in the Montauk Project. Montauk, also known as the Phoenix Project, used Bielek and his brother Duncan Cameron, to explore the underground cities of Mars. SS: Sovereign Scribe – AB: Al Bielek SS: Regarding your experience on Mars you walked through the time tunnel, you take a step and you're on Mars: What did you see? AB: Well I was not... Read More

Anyway, another incredible example of all this in the Cymatics videos is seeing almost:human-like figures forming from the particles when certain sounds are emitted. Our bodies are also the result of sound resonating energy into form and if our minds are powerful enough to change the sound range of the body, it moves into another form or disappears from this dimension, altogether. This is what is called shape-shifting. It is not a miracle, it is science, the natural laws of creation. The full-blood reptilians of the lower fourth dimension can therefore make their 'human' physical: form disappear and ~ bring forward their reptilian level of existence. They shape-shift. To us in this dimension they appear human, but it's just a vibrational overcoat, Hillary Clinton appeared as a reptile, while her husband, Bill Clinton the US President, was only overshadowed, and controlled by one. This is interesting because my own research, and: that of others, has revealed Hillary Clinton to be much higher in the hierarchy than Bill, who while of a crossbreed bloodline, is a pawn in the game, to be used and discarded as necessary, It is not always that the most powerful people are placed in what appears to be the most powerful jobs.

This reminds me of yet another account from the early 1980's which was covered in a now-defunct newsletter called "THE CRYSTAL BALL", which printed a "Shaver Mystery" special edition. This edition recounted a story of Russian scientists who, searching for a meteorite impact in northern Siberia, discovered an underground facility beneath the ice and snow. Several humanoid bodies were discovered in frozen suspended animation. Some of these were revived but they proved to be far from human, but rather reptilians who had somehow shape-shifted into human form... whereas the true humans in this ancient scientific colony could not be revived. Eventually the whole scientific team was "assimilated" by the reptilians, the reptilians somehow reaching into their minds and absorbing their memories and physical features. Other scientists were called in and these in turn were assimilated/absorbed by other reptilians. The story has it that the infiltration reached even to the deepest levels of the Russian Politbureau and if true it may have paved the way for other "infiltratration" agendas involving other countries.

Between the Archuleta Mesa of New Mexico (the main headquarters of the malevolent Reptiloid forces) and Death Valley in California (below which lies the main headquarters for the benevolent humanoid forces) there are several bases where things are "out of control". These consist of huge cavern systems linked together via artificial tunnels… with main "bases" below Deep Springs, CA; Mercury, NV; Dougway SW of and Granite Mt. SE of Salt Lake City; Page, AZ; Creede and also the Denver International Airport in Colorado, etc.

The benevolent humanoid Federation forces involved are from the Andromeda and Pleiades constellations, and also Tau Ceti, Vega Lyra, Procyon, Wolf 424, Alpha Centauri, etc. The malevolent Reptiloid Empire forces are from Draconis and Orion, and also Epsilon Bootes, Zeta Reticuli II, Capella, Polaris, etc.

etween 1979 when the Dulce and Groom wars broke out leading to the take-over of "our" joint operational bases and 1989 when the reptiloids/ grays took control of the Alternative-3 bases on Luna and Mars, several of the Melchizedek bases were also attacked as the reptiloids/grays turned on these native subterranean residents. During the two-year period when the executive branch of our government broke relations with the Grays following the Dulce-Groom wars, the intelligence community split into two factions: the American-Navy backed COM-12 agency which no longer desires interaction with the Grays, but seeks to maintain contact with the Pleiadeans instead and is fighting to preserve Constitutional government; and the Bavarian-CIA backed AQUARIUS agency which seeks to maintain contact with the Grays, etc. in that they are depending on their mind control technology, abductions and implants to impose a joint human-alien fascist "New World Order" dictatorship.

. This is an important distinction. There are the 'full-bloods' who are reptilians using an apparent human form to hide their true nature, and the 'hybrids', the reptile-human crossbreed bloodlines, who are possessed by the reptilians from the fourth dimension. A third type are the reptilians who directly manifest in this dimension, but can't hold that state indefinitely. Some of the 'Men in Black' are examples of this.

Dear ;

I am an abductee who has experienced suppressed encounters with grays and humans throughout an underground system that spans the western base of the Wasatch-Rocky mountains in Utah.

Alien power plant (Photo credit: wili_hybrid)

I've had "altered state" contacts with the humans who live within the underground system, those who I refer to as the "Melchizedeks". These are members of a metaphysical lodge with connections to the deep initiatory levels of Mormonism, Masonry, the Mt. Shasta/Agharti network, Mayans, Sirius, Arcturus, Saturn, etc. They had formerly maintained cautious territorial treaties with the branch of grays/reptiloids that are native to the underground levels, and the reptiloids took advantage of this agreement in order to infiltrate their society.

Then there are the experiences of Cathy O'Brien, the mind controlled slave of the United States government for more than 25 years, which she details in her astonishing book, Trance Formation Of America, written with Mark Phillips. She was sexually abused as a child and as an adult by a stream of famous people named in her book. Among them were the US Presidents, Gerald Ford, Bill Clinton and, most appallingly, George Bush, a: major player in the Brotherhood, as my books and others have long exposed. It was Bush, a pedophile and serial killer, who regularly abused and raped Cathy's daughter, Kelly O'Brien, as a toddler before her mother's courageous exposure of these staggering events forced the authorities to remove Kelly from the mind control programme known as Project Monarch. Cathy writes in Trance Formation Of America of how George Bush was sitting in front of her in his office in Washington DC when, he opened a book at a page depicting "lizard-like aliens fro m a far off, deep space place." Bush then claimed to be an 'alien' himself and appeared, before her eyes, to transform 'like a chameleon' into a reptile. Cathy believed that some kind of hologram had been activated to achieve this and from her understanding at the time I can see why she rationalised her experience in this way. Anyone would, because the truth is too fantastic to comprehend until you see the build up of evidence. 'There's no doubt that alien~based mind programmes are part of these mind control projects and that the whole UFO-extraterrestrial scene is being massively manipulated, not least through Hollywood films designed to mould public thinking. Cathy says in her book that George Lucas, the producer of Star Wars, is an operative with NASA; and the National Security Agency, the 'parent' body of the CIA." But given the evidence presented by so many other people, I don't believe that what Bush said and Cathy saw was just a mind control programme. I think he was revealing the Biggest Secret, that a reptilian race from another dimension has been controlling the planet for thousands of years. I know other people who have seen Bush shape-shift into a reptilian.

. It seemed as if the reptilian tongue could not pronounce the word "KIN-IN-I-GIN". If a suspected reptilian infiltrator was cornered and could not bring itself to pronouncing the words, they were taken and if proven to be reptilian they were dealt with accordingly.

. It seemed as if the reptilian tongue could not pronounce the word "KIN-IN-I-GIN". If a suspected reptilian infiltrator was cornered and could not bring itself to pronouncing the words, they were taken and if proven to be reptilian they were dealt with accordingly.

The president of Mexico in the 1980s, Miguel DeLa Madrid; also used Cathy in her mind controlled state. She said he told her the legend of the Iguana and explained that lizard-like extraterrestrials had descended upon the Mayans in Mexico. The Mayan pyramids, their advanced astronomical technology and ~ the sacrifice of virgins, was inspired by lizard-like aliens, he told her."' He added that these reptilians interbred with the Mayans to produce a form of life they could inhabit. De La Madrid told Cathy that these reptile-human bloodlines could, fluctuate between a human and iguana appearance through chameleon-like abilities – "a perfect vehicle for transforming into world leaders", he said. De la Madrid claimed to have Mayan-lizard ancestry in his blood which allowed him to transform back to an iguana at will. He then changed before her eyes, as Bush had, and appeared to have a lizard-like tongue and eyes." Cathy understandably believed this to be another holographic projection, but was it really? Or was De La Madrid saying something very close to the truth? This theme of being like a chameleon is merely another term for 'shape-shifting', a theme you find throughout the ancient world and among open minded people, in the modern one too.

Contactee Maurice Doreal may add something new to all of this, with his claim that he was invited — by two "blond men" who attended one of his lectures — into an ancient neo-Mayan city under Mt. Shasta, California called Telos [interesting enough, "Telos" is also a Greek word meaning "uttermost" or "purpose"]. During later contacts Doreal was shown some ancient 'holographic' libraries beneath the Himalayas, and holographic records of a technically-advanced race of tall, blond and blue-eyed humans who ruled a vast empire where the Gobi desert now lies.

These 'Nordics' were at war with a race of reptilian or neo-saurian humanoids — velociraptor type humanoids, possibly the result of ancient genetic engineering gone out of control? — based on what at the time was the semi-tropical continent of Antarctica. The 'Nordics' literally drove the reptilian humanoids off the face of the earth, the reptiloids taking refuge in vast underground cavern systems [possibly akin to "Snakeworld", "Patala" or "Nagaloka" with it's reptilian capital "Bhoga-vita" — which is according to Hindu tradition part of a seven leveled subterranean realm stretching from Benares India to Lake Manosarowar Tibet, and inhabited by deadly reptilian humanoids called the 'Nagas']. There they developed a hive-like society in order to advanced their occult-technology.

[others have suggested that many native Americans have a specific meta-gene factor that could potentially prove to be very threatening to the reptilian agenda, which is why the "lizard people" closely monitor native Americans and attempt to keep them under their hypnotic "spell"].

01:17:00 TOOL LATERALUS TALK ABOUT 51 AREA ....TO GET TO "NEXUS"

Fortunately the Luciferian collectivists according to Revelation 12 will lose their power base in the galaxy as resistance to their atrocities increases. Unfortunately however the central command of the collective will escape to the caverns of planet earth, which will serve as their

"last stand", and according to prophecy they will begin a desperate program to recruit Terrans through a European-based New World Order involving electronic mind control implants and ancient roots in the remnants of the ill-named Holy Roman Empire. All of this will be a last-ditch effort to re-gain their lost ground among the stars. In the process they will devastate much of the planet, but it will all be for naught as they will lose in the end. The question is NOT whether they will lose the war, the question is how many of US will survive these apocalyptic events. I believe that this largely depends on the collective WILL of individual human beings throughout this planet. It is not something that is set in stone.
-Branton

#LA CLAVE ESTA EN VENUS COMO SABEMOS EN 4 O 5 O 6 O 7 DIMENSION...MIRAR BARCO A VENUS LA CANCION ESCUCHARLA DE MECANO...QUE ES UN MECANO?...HH.

#RUTA DE ACCESO A NOTAS DE "NEXUS" EN EL CARPESANO CON CLIP GIGANTE DE LA PARTE DE ARRIBA SUPERIOR IZQUIERDA ENFRENTE AL ENTRAR AL HOSPICIO...CENTRO MODULAR DE LA GALAXIA EPSYLON...HH RUTA DE ACCESO...HH.

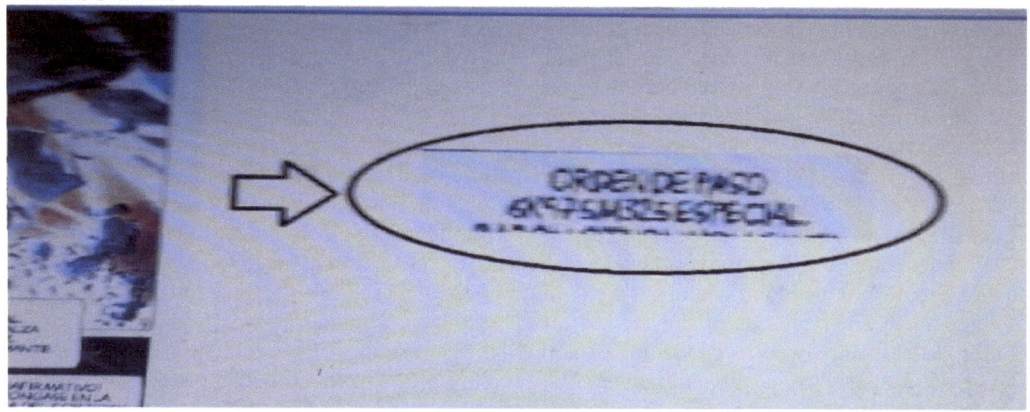

ORDEN DE PASO 6K975M325 ESPECIAL...HHH

"Bonnie, her mother (Rana Mu), her fatherRa(Mu), her sister Judy, her cousins Lorae and Matox, live and move in our society, returning frequently to TELOS for rest and recuperation. Bonnie relates that her people use boring machines to bore tunnels in the Earth. These boring machines heat the rock to incandescence, then vitrify it, thus eliminating the need for beams and supports. A tube transit tunnel is used to connect the... cities that exist in various subterranean regions in our hemisphere. The tube trains are propelled by electromagnetic impulses up to speeds of 2500 mph. One tube connects with one of their cities in the Matto Grosso jungle of Brazil. (They) have developed space travel and some flying saucers come from their subterranean bases...

"They grow food hydroponically under full-spectrum lights with their gardens attended by automatons. The food and resources of Telosare distributed in plenty to the million-and-a-half population that thrives on a no-money economy. Bonnie talks about history, of theUighers, Naga-Mayas, and Quetzals, of which she is a descendant

(Note: Many people have mistakenly identified the inhabitants of 'Telos' as being directly descended from the 'Lemurians', however Bonnie here seems to refute this by indicating that her ancestrage was other than this, possibly Meso-American and/or East-Indian? As in the case of the ancient 'antediluvian' cities of the eastern seaboard which were re-established after being abandoned by the lost 'Atlanteans', the 'Lemurians', if they existed, also seem to have been devastated in a world-wide cataclysm and their cities re-established by the Uighers, Naga-Mayas, and Quetzals and probably scattered members of other societies. As we've said earlier, the name 'Telos' is a Grecian word meaning 'uttermost', suggesting a 'possible' connection with the grecian-like Hav-musuvs of the Panamint mountains of California - Branton).

"I met Bonnie's cousin, Matox, who, like her, is a strict vegetarian and holds the same attitudes concerning the motives of government. They constantly guard against discovery or intrusion. Their advanced awareness and technology helps them remain vigilant...

"Science Fiction? Bonnie is a real person. Many have met her. Is she perpetrating a hoax? For what motive? She does not seek publicity and I have a devil of a time getting her to meetings to talk with others, but she has done so. There has been little variation in her story and her answers in the past three years. She has given me excellent technical insight on the construction of a crystal-powered generator that extracts ambient energy... Bonnie's father, the Ramu, is 300 years old and a member of the ruling council of Telos.

"Many tunnels are unsafe and closed off. All tube transit tunnels are protected and are designed to eject uninvited guests. Does Bonnie have the answers that we are looking for? I don't know... Bonnie says she would like to satisfy our need for proof and will work with me on a satisfactory answer to that problem, but she is unconcerned with whether people accept her or not. Bonnie is humorous and easy-going and well-poised, yet sometimes she becomes brooding and mysterious. She says her people are busy planning survival centers for refugees. One of these is to be near Prescott, Arizona..."

(Note: or rather below the Groom Creek area just south of Prescott, to be exact. Another 'survival center' for refugees of the world-wide cataclysms which the Telosians believe will eventually devastate the surface of the earth, is said to be below the general area of Jenny Lake, Wyoming, near the Tetons. The Tetons themselves have been the alleged home of a mysterious race, according to different sources, and extremely ancient stone 'buildings' have reportedly been found high atop these peaks - Branton).

When Bill Hamilton asked "Bonnie" to elaborate about the power- sources which her people utilize to propel the so-called "flying saucer" craft, she replied:

"...A lot of it is crystals (i.e. crystal-induced electromagnetism? - Branton), particularly the atmos

01:17:00 TOOL LATERALUS TALK ABOUT 51 AREA ....TO GET TO "NEXUS"

"Still another explorer named D.O. visited this same tunnel near Gaspar, Santa Catarina, and behind a wonderful fruit orchard saw a subterranean woman with a child in her arms reading to it aloud from a huge book written in an unknown language... After she read each sentence the child repeated the same and in this way was taught how to read. All of these subterranean cities are illuminated by strange light...'"

In relation to the apparent connection between subterranean civilizations and unidentified flying objects (Bernard and de Souza, incidentally, believed 'flying saucers' to be of subterranean origin), we will here quote from Paris Flammonde, author of 'THE AGE OF FLYING SAUCERS' (Hawthorne Books, Inc., N.Y.), who tends to confirm this hypothesis. He in turn quoted Raymond A. Palmer as a major proponent of this belief:

"...The new decade was not without a new theory, or, at least, a variation of an old one--that not only were Flying Saucers not originating from beyond the farthest reaches of our planet, they were expelled from within it...Ray Palmer wrote a lengthy article elaborating his interesting and imaginative thesis, and prefaced it with the assertion that he was prepared 'to prove that flying saucers are native to planet earth; that the governments of more than one nation (if not all of them) know this to be the fact; that a concerted effort is being made to learn all about them, and to explore their native land; and that facts already known are considered so important that they are the world's top secret...' The continuation of his contention reads:

'...is there any area on Earth which can be regarded as a possible origin for flying saucers? There are... four... the two major, in order of importance, are Antarctica andthe Arctic... the two minor areas are South America's Motto Grosso and Asia'sTibetan Highlands.'"

Raymond Bernard (actual name 'Walter Seigmeister'), writing in the Oct. 1959 issue of SEARCH Magazine, p. 48, described yet another alleged encounter with a subterranean race. What are we to make of all these stories? Are we to assume that some of the individuals who toldBernard such accounts actually made them up, as some suggest, in order to receive the 'reward' Bernard was known to offer on documentable accounts of ancient tunnels? Or, are we to accept these accounts for just what their sources claim them to be, actual encounters with a subterranean world? Bernard stated the following:

"...Last week my investigators returned and said they visited their city (i.e. the 'city' of a race of dwarf-humans whom Bernard referred to as the 'Niebelungs', who live in a subterranean region with it's own system of illumination - Branton) and are able to bring any of my American friends to visit it, but I require one condition: absolute secrecy, as I don't want governments to send armies into the tunnel to disturb these peaceful people.

"To reach them requires a 3-day journey of about 40 miles through a tunnel. This entire distance is through a tunnel carefully lined with cut stone blocks below, above and on the sides. That was quite an engineering feat. I think the tunnel was made long to keep out curiosity seekers, and only the most determined will travel that distance.

"Here is the report of my investigations: (They are two ranchers, father and son, who discovered the tunnel accidentally):

"'We left our house 5 A.M. for the tunnel on top of a mountain and reached it 3 P.M. We were tired and camped near the entrance of the tunnel. For three days we proceeded through the tunnel. We told time by our watches, as we could not tell when it was day or night. We went to sleep at 10 P.M. and awoke at 3 A.M. and continued walking. By the third day the tunnel started to go downward by steps. It was built of stone blocks on all sides. By the night of the third day the tunnel suddenly opened into a great space covered with what appeared as a sky with a yellow light that made everything luminous, like daylight. We saw a city with many houses and saw many people in the distance. They were dwarfs with long

#TÚ ERES LA NOCHE ETERNA...Y SOLO SE RECUBRIO DE CARNE DE LABIOS DE UÑAS DE PEOR SI ABRIESEMOS PERO POR TUS OJOS Y TÚ CABELLO EL TIEMPO NO HA PODIDO CUBRIR Y SALE MOSTRANDOSE ..Y SI ABRIESEMOS PIR DENTRO DE TI NO HABRIS MUSCULOS HUESOS SANGRECSINO LAVPROPIA NOCHE OSCURA Y PROFUNDA LS ETERNIDAD DEL COSMOS HH...Escribir cuento..hh.
un lenguaje mixto entre yo y branton perfect!.
BAJAR METABARONES Y VER LOS COMICS EN EL JARDIN PARA REDACTAR "NEXUS" MUY EN EL CONTEXTO DE LA VALENCIA DE 1984 CONTEXTUALUZ EL LIBRO AHI Y EN LA VALENCIA D PEREZ CASADO PERO SIN PERDERTE EN LOS DETALLES HISTORICOS SINO EN LA PSICOLOGIA DE LOS PERDONSJES MAS COMO UNA EXCUSA QUE COMO UNA INTROD. HISTORICA Y HAWAIKA...COMICS METAL HURLANT...Y ESCRITO EN BRASIL...COMO AHORA EXORZIZAR LOS MOV REPTILISNOS DE AHORS MISMAS ARMAS METAPSICOLOGICAS...AH Y LO MAS IMPORRANTE RECORDAR QUE ESTAMOS EN UN AÑO ABEJA-ARTE (GAY) Y QUE ES UN AÑO NO PARA CAMBIAR NADA SINO PARA CREAR..Y LEER OTRA VEZ "EL GRAN PLAN DIVINO" OBLIGAO!!!... 😊😊😊😊 .SIN LIMITACIONES...HH..PLAZA DE OTROS ESTACION DEL NORTE MALVAROSA DISEÑADOR DE VALENCIA MONTESINOS DISCOTECA DE LA MALVAROSA...NOMBRE...HH PERO VOLVER AQUI PARA VOLVER OTRA VEZ ALLI Y CAMBIARLO TODO LA HISTORIA...HH Y DE NUEVO VUELTA AHORA A LA GALAXIA LYRA CON XENUS....Y BRASIL 2015 EN RIO GRANDE DO SUL ESA TEXTURA...HH...METER PROTEUS EN VALENCIA...HH...EL AÑO 1981 FUE OTRO AÑO DE METAL...HH EL AÑO K VIENE SI SERA UN AÑO DE VIAJES MUCHOS VIAJES...HH.
jEXPERIENCIAS GENETICAS REPTILIANAS...NEXUS...FOTO...#geneticalreptiliansxxxc en LENOVO.Hum.
Faltas muy graves
(Entre 30.001 y 600.000 euros de sanción)Manifestaciones no comunicadas o prohibidas ante infraestructuras críticas.Fabricar, almacenar o usar armas o explosivos incumpliendo la normativa o careciendo de la autorización necesaria o excediendo los límites autorizados.Celebrar espectáculos públicos quebrantando la prohibición ordenada por la autoridad correspondiente por razones de seguridad pública.
Some people, as strange as it may sound, believe that there is a conspiracy in effect upon and beneath planet earth, one that is designed to slowly and subtly enslave us through a constant

barrage of subliminal programming, economic manipulation, and preconditioning. This plan is one involving a scenario that is designed to set all countries, nations or republics against each other [eventually doing away with all sovereignties altogether and replacing them with a global religio-eco-political control system], using infiltrators who operate within the leadership ranks of all countries, especially the USA, Europe, Russia, Red China, Australia and so on. In other words our planet is the chessboard, the countries are the squares, we are the 'pawns', and the draco are the chess players. If this conspiracy continues as planned they could "divide-and-conquer" us into oblivion, with MOST of the human population being eliminated (wars, famine, plagues, etc.). THEN they plan to emerge from their underground empire consisting of multi-connected bases — Dulce, Pine Gap, Gizeh, Neu Schwabia, etc., where most of the humans have since been.
#metabarones1anexusxxxxx
#metabarones2anexusxxxxx
dos archivos de JPEG del ordenador para meterlos en el libro"NEXUS"...HH

#EL CENTRO MODULAR DE LA GALAXIA EPSYLON NO SOLO ESTÁ EN COMUNICACION CON LA CIUDAD SUBTERRANEA DE LOS INTEARERRESRES SINO DE UNAS 5 O 6 CIUDADES AQUI DEBAJO DE ESTALAGEM...HH...Y YO TEBGO LA VENTAJA DE ESTAR EN COMUNICACION CON TODAS ELLAS QUE COMO SE HACE?SUMPLEMENRE SABIENDO DE ELLAS YA ELLAS TE PONEN EN CONTACTO CON SUS MAQUINAS Y TECNOLOGIAS Y TE VISIONAN TELEVISIONES DE PLASMA GIGANTES Y TE SUGUENN PASO A PASO CON LO K HACES Y SI ESTAS HACIENDO COSAS EN INTERNET SOBRE ELLOS...PORQUE SU PRINCIPAL TRABAJO SOMOS NOSOTROS...HH...LOS QUE PRETENDEN VIVIR UNA VIDA NORMAL CONSENSUAL SON EN REALIDAD SONAMBULOS Y ZOMBIES...A MERCED DE LAS ONDAS DE TRACCION DEL REPTILUANO MAYOR EN UNA DE ESAS CIUDADES SUB ESTALAGEM O SUS SECUACES Y AYUDANTES EN LA SUPERFICIE...HH.
til the 1950's."
Note: There are indications that some members of certain Masonic- type 'secret government' societies, such as the Rosicrucian Order, have attempted to establish contact with the subterranean residents of Mt. Shasta, although it is uncertain just what might have come of this. Several encounters with the 'Blondes' (both subterran and exterran?) have revealed 'their' own concern about what is taking place with the abductions and mutilations of human beings by the sauroid Grays, although many of these groups claim that they cannot 'interfere' with the problem due to some 'cosmic law' of non- intervention. This may be true with those 'Nordic' or 'Blonde' societies who hail from other planetary bodies, such as the Taurians, Lyrans, Eridanians, and Cetusians (the latter of whom seem to be taking the most action to help their brothers here on earth, in essence interfering with the saurian 'interferers' from the Draconis, Bootes, Reticuli, Canis, etc. constellations), and the 'Solar Tribunal' groups of Mars, Luna, Saturn, etc, and so on. However, in the case of the Telosian-Aghartian alliance, this 'non-intervention' policy would not

apply, since this is their world also, and they are just as native to earth as anyone else living on this planet. In light of this fact, and especially in light of their own awareness of the reptilian-saurian threat, we would urge them (if by chance they are reading this) to reconsider such a stance and join with their fellow human brothers and sisters on the surface in defending our society from this ancient threat

Others suggest that not only in southern Utah near Page, Arizona but also in northern Utah near Dugway [Page and Dugway according to former Dulce base security officer Thomas E. Castello being the two MAJOR underground connections between the DULCE New Mexico base and the GROOM Lake Nevada base], massive societal infiltration of reptilian entities in human guise has occurred. These reports are widespread, however these "chameleoids" have been seen more profusely near Dulce New Mexico, Dugway Utah, and Area 51 Nevada. There are also rumors of draconian infiltration of the federal and state government agencies, military facilities, mental health facilities, religious organization, industrial complex, and even high levels of the police forces in the state of Utah by especially the subterraneous counterparts of the "reptilian" species who exist EN MASSE within the underground levels beneath the state, since the connections between the Dulce and Groom bases not only involve artificial tunnel passages, but also an ancient and huge natural cavern network which the "aliens" have reportedly "stolen" from native American sub-colonies in years past in remarkably similar fashion as the early "Shaverian" accounts of subterranean Tero/Dero conflicts… the final takeover of these cavern-colonies or tribes coinciding with the outbreak of the Dulce and Groom Wars which raged from between 1979 to 1985 and resulted in many of "our" joint operational multi-trillion dollar underground facilities being taken over by the Dracos, their "Grey" subordinates, and in some cases insectoid collaborators [more of our tax dollars disappearing "down the tubes" to feed these vamperial parasites]…

#EXOBIOLOGIA...EXISTE EL HOMBRE LOS ANIMALES LAS PLANTAS Y LA EXOBIOLOGIA O SERES BIOLOGICOS ALIENS AQUI EN NUESTRO PLANETA..DE HECHO LLEVAN AQUI MAS TIEMPO QUE NOSOTROS CON LO CUAL NOSOTROS SERIAMOS LOS ALIENS Y/O EXTRATERRESTRES..O UNA RAZA MAS AHORA EN PROCESO DE AUTODETERMINACION DEL RESTO DE RAZAS SIMBIOTICAS : :
#RUTA DE ACCESO A NOTAS DE "NEXUS" EN EL CARPESANO CON CLIP GIGANTE DE LA PARTE DE ARRIBA SUPERIOR IZQUIERDA ENFRENTE AL ENTRAR AL HOSPICIO...CENTRO MODULAR DE LA GALAXIA EPSYLON...HH RUTA DE ACCESO...HH.

Once we are able to access the technology that the 'elite' have STOLAN from us then we can finally break out of this 'cradle' and our energy, population, pollution, and economic problems will cease… however we should and must of course refrain from violating the sovereignty of other worlds in doing so.

#YO ENCONTRE EL SIGNIFICADO DEL UNIVERSO UNA TARDE DE VERANO HABLANDO CON JUAN ,MI ABUELO, SENTADOS SOBRE UNOS 15 SACOS DE CEBADA...MI ABUELO METIO SU MANO GRANDE Y ARRUGADA CASI NEGRA DEL SOL PERTINAZ Y EL TRABAJO INCOLUME Y SACO UN MONTON DE DENTRO: "AQUI ESTA EL SENTIDO DE TODO" Y DEJO QUE LOS GRANOS DE LA CEBADA RECIEN COSECHADA VOLVIERAN AL INTETIOR DEL SACO DE TELA...COMO EL AGUA FRESCA DE UN MNANTIAL..."Y LOS HOMBRES SOMOS LOS GRANOS DEL SACO"SIEMPRE INTENTAMOS DESTSCAR DEL RESTO DE GRANOS PERO AL FINAL TODOS VOLVEMOS AL INTERIOR DEL SACO SIENDO OTRS VEZ TODOS LOS GRANOS LO MISMO" ME DIJO CON UNA AMPLIA SONRISA Y DANDO UNA SUAVE PERO CONTUNDENTE PALMADA EN EL AIRE PARA LIMPIARSE LOS RESTOS DEL POLVO QUE DEJABAN LOS GRANOS..M" AL FINAL NO QUEDA DE NOSOTROS NI EL POLVO"VOLVEMOS AL ORIGEN.
Y ASI DE ESA FORMA ME EXPLICO TODO:EL ORIGEN DE LA VIDA LOS MISTETIOS DE LA VIDA Y LA MUERTE Y LA COMPOSICION DEL UNIVERSO Y SU DESTINO...AÑOS DESPUES PUDE CERTIFICAR TODA SU COSMOVISION ...Y QUE COINCIDIA COMPLETAMENTE CON LAS TEORIAS MAS DESARROLLADAS DE LA FISICA CUANTICA Y DE KA BIOTECNOLOGIA MAS ARRIESGADAS Y CON EL

SIMIL ENLAZABA A LA VEZ CON TODAS LAS "LEYENDAS" DE LOS LIBROS DEL HINDUISMO EL "MAHABARATA"...NO EN VANO LOS APORTES DE LOS GURUS DE LA INDIA DURANTE LA EPOCA NAZARITA EN GRANADA DEJARON HONDS HUELLA EN AQYELLAS TIERRAS...MUY PROFUNDO QUIEN SABE POR APORTES DE OTROS PUEBLOS ANTERIORES QUE POBLARON AQUELLAS TIERRAS Y HABIAN ESTADO EMPARENTADAS CON LA INDIA O CON LOS POBLADORES ORIGINALES EXTRATERRESTRES DE ESTAS Y AQUELLAS TIERRAS (INCLUIR EN EL LIBRO "NEXUS")...HEHE!!!

Y LE PREGUNTARON AL SABIO : CUAL ES EL SENTIDO DEL UNIVERSO ..Y EL SABIO SENTADO SOBRE UN SACO DE CEBADA EN AQUELLA TORRIDA TARDE DE VERANO LES MIRO CON UNA SONRISA AMPLIA EN LOS LABIOS : CUANDO COMPRENDAIS LO QUE MUESTRA LA MIRADA DE UN GATO VOLVED Y HACEDME LA PREGUNTA ¡...TODOS SE FUERON Y AÑOS DESPUES TODOS REGRESARON ..ALLA DOND ESTABA EL SACO DE CEBADA Y EL SABIO SENTADO SE HABIA TRANSFORMADO EN UN CAMPO DE CEBADA QUE LLEGABA HASTA EL MAR Y QUE DABA DE COMER A MILLONES DE PERSONAS ..ENTONCES COMPRENDIERON Y SE EXTENDIERON POR TODOS LOS RINCONES DEL MUNDO "..ESE SABIO ERA MI ABUELO,JUAN..A EL LE DEDICO ESTE CUENTO..THE RESISTANE GOES ON´2015!!!...HEHE!!!>

Another man, J.B., claims to be a nocturnal astral vampire-killer, part of a team of nocturnal astral/dream warriors, a type of astral special forces who have declared war on the fifth density reptilian life-force vampires — the very same fifth density dracos who have reportedly taken over powerful world leaders, having assimilated their 'hosts' into their own beings which manifest in true reptilian form as some type of "wer-drac" manifestation during their secret "blood-fests". J.B. later confirmed the claims of K.S. when he stated that reptilian infiltrators [or chameleoids – both the 5th density "snatchers" and the 3rd density "shifters"] favored a particular shopping center in Salt Lake City. Some of these fifth density snatchers may actually have gained their third density solidity by fully assimilating/absorbing their "hosts" in similar fashion as the legendary "wer" people, thus becoming third density "shifters". Actually, all three types may exist, the repti-poltergeist parasites or "SNATCHERS"; actual third density and most likely subterraneous reptiloid "SHIFTERS" – using laser holograms, technotic projection, superficial bio-phasing, etc., to accomplish the shifts; and the intermediary entities or "SNATCHER/SHIFTERS"!?

The Underworld Empire
Part 4
Pages | 1 | 2 | 3 | 4 | 5 | 6 | 7 |

Regardless of "Joe's" opinion, however, there is reason to believe that influences from these nether regions can and do affect "us" in a profound way, and even the men whom Ralph and Joeencountered, whoever they were, admitted this fact.

Is there anything else which we might be able to "read into" this scenario, based on the accumulated data which we've given in previous files? The men who were encountered do confirm than an ancient (antediluvian?) race did in fact leave behind extremely sophisticated technology, and it is probably true that man in his largely unregenerate state might be influenced to destroy themselves with these sophisticated machinery if given the chance. Then again the so- called Horlocks(perhaps the same as the 'cybernized', mind-altered and controlled "Men In Black" described by John Keel and others!?) have seemingly utilized such technology without utterly destroying themselves. This could be due to the fact that their 'controllers' (the serpent races?) realize the dangers of such technology and desire to conquer without destroying that which they are conquering.

Also, man already has enough 'technology' in the form of nuclear weaponry, etc., to destroy himself many times over, but no use adding fuel to the fire as they say. As for these underground or subsurface people, they are apparently part of a race or races who discovered these recesses either hundreds or thousands of years ago, or perhaps different groups who discovered this network throughout this entire period of time. The 'horlocks' seem to be a group working under an evil influence, for instance--as we've said--possibly that of the serpent race, since there have been documented CONNECTIONS uncovered between the MIB and theSerpent Race as we have seen

I GREW UP AT LYRA WITH MAMOUTH'S MILK STRONG AND SMART...new book.
"The entrance to Golondrinas is located in one of the most primitive and uncivilized areas of Mexico and local inhabitants are afraid to approach the cave because they believe it is full of 'evil spirits' which lure people to their deaths. They tell stories of people mysteriously disappearing never to be heard from again while passing near the cave entrance. These stories may be based more on fact than fiction: they are similar in some respects to UFO abduction reports. Because of its huge size, remote location, and unique geological structure, Golondrinas would be an ideal UFO base. Naturally camouflaged cBut some things that CAC refers to seem to make sense, especially the following quote in response to reports that an alien conflict resulted in a planetoid and accompanying ships crashing into Jupiter. This planetoid was reportedly arriving from a dark star outside of our system called Nemesis, in the direction of Orion — a sphere which if it had enough mass would have become our nearest stellar neighbor, yet lacking mass it condensed into a large frozen planet about the size of Jupiter. A DARK star. This planetoid was reportedly filled with 40 million Draco warriors in cryogenic freeze. The loss of this armada reportedly set back the Draco/New World Order takeover plans for planet earth considerably. It could be that this report was completely false, but even if so there is still much evidence suggesting an alien connection to the New World Order nonetheless. CAC stated:

"This Awareness indicates that this is a great setback for the New World Order, for the New World Order was a plan BY AND FOR the Reptoids, and it would have benefited greatly had the Reptoids made their invasion on earth. It would have led to the need for a New World Order…"

It makes sense. Unite all nations into a central power and you will only have to take control of a few key individuals rather than dealing with numerous stubbornly independent sovereignties.
aves in other parts of the world may serve as excellent natural bases, way stations, and 'depots' for UFOs.

http://www.subterraneanbases.com/cave-and-tunnel-entrances-of-south-america/

Did the Philadelphia Experiment open up more than just a hole in hyperspace? Could it have actually opened up a doorway for other-dimensional 'Dracos' to come flooding into our world in massive numbers, even to the point of infiltrating our world and replacing key political figures in order to impose their agenda upon us? Read the following and you might just begin to wonder.

Proyectar haces de luz sobre los pilotos o conductores de medios de transporte que puedan deslumbrarles o distraer su atención y provocar accidentes.

Faltas graves

Entre 601 y 30.000 euros de multaPerturbar la seguridad ciudadana en actos públicos, espectáculos deportivos o culturales, solemnidades y oficios religiosos u otras reuniones a las que asistan numerosas personas.La perturbación grave de la seguridad ciudadana en manifestaciones frente al Congreso, el Senado y asambleas autonómicas aunque no estuvieran reunidas.Causar desórdenes en la calle u obstaculizarla con barricadas.

#LA CLAVE ESTA EN VENUS COMO SABEMOS EN 4 O 5 O 6 O 7 DIMENSION...MIRAR BARCO A VENUS LA CANCION ESCUCHARLA DE MECANO...QUE ES UN MECANO?...HH

#YO ENCONTRE EL SIGNIFICADO DEL UNIVERSO UNA TARDE DE VERANO HABLANDO CON JUAN ,MI ABUELO, SENTADOS SOBRE UNOS 15 SACOS DE CEBADA...MI ABUELO METIO SU MANO GRANDE Y ARRUGADA CASI NEGRA DEL SOL PERTINAZ Y EL TRABAJO INCOLUME Y SACO UN MONTON DE DENTRO: "AQUI ESTA EL SENTIDO DE TODO" Y DEJO QUE LOS GRANOS DE LA CENADA RECIEN COSECHADA VOLVIERAN AL INTETIOR DEL SACO DE TELA...COMO EL AGUA FRESCA DE UN MNANTIAL..."Y LOS HOMBRES SOMOS LOS GRANOS DEL SACO"SIEMPRE INTENTAMOS DESTSCAR DEL RESTO DE GRANOS PERO AL FINAL TODOS VOLVEMOS AL INTERIOR DEL SACO SIENDO OTRS VEZ TODOS LOS GRANOS LO MISMO" ME DIJO CON UNA AMPLIA SONRISA Y DANDO UNA SUAVE PERO CONTUNDENTE PALMADA EN EL AIRE PARA LIMPIARSE LOS RESTOS DEL POLVO QUE DEJABAN LOS GRANOS..M" AL FINAL NO QUEDA DE NOSOTROS NI EL POLVO"VOLVEMOS AL ORIGEN.

Y ASI DE ESA FORMA ME EXPLICO TODO:EL ORIGEN DE LA VIDA LOS MISTETIOS DE LA VIDA Y LA MUERTE Y LA COMPOSICION DEL UNIVERSO Y SU DESTINO...AÑOS DESPUES PUDE CERTIFICAR TODA SU COSMOVISION ...Y QUE COINCIDIA COMPLETAMENTE CON LAS TEORIAS MAS DESARROLLADAS DE LA FISICA CUANTICA Y DE KA BIOTECNOLOGIA MAS ARRIESGADAS Y CON EL SIMIL ENLAZABA A LA VEZ CON TODAS LAS "LEYENDAS" DE LOS LIBROS DEL HINDUISMO EL "MAHABARATA"...NO EN VANO LOS APORTES DE LOS GURUS DE LA INDIA DURANTE LA EPOCA NAZARITA EN GRANADA DEJARON HONDS HUELLA EN AQYELLAS TIERRAS...MUY PROFUNDO QUIEN SABE POR APORTES DE OTROS PUEBLOS ANTERIORES QUE POBLARON AQUELLAS TIERRAS Y HABIAN ESTADO EMPARENTADAS CON LA INDIA O CON LOS POBLADORES ORIGINALES EXTRATERRESTRES DE ESTAS Y AQUELLAS TIERRAS (INCLUIR EN EL LIBRO "NEXUS")...HEHE!!!

#EL CENTRO MIDULAR DE LA GALAXIA EPSYLON NO SOLO ESTÁ EN COMUNICACION CON LA CIUDAD SUBTERRANEA DE LOS INTEARERRESRES SINO DE UNAS 5 O 6 CIUDADES AQUI DEBAJO DE ESTALAGEM...HH...Y YO TEBGO LA VENTAJA DE ESTAR EN COMUNICACION CON TODAS ELLAS QUE COMO SE HACE?SUMPLEMENRE SABIENDO DE ELLAS YA ELLAS TE PONEN EN CONTACTO CON SUS MAQUINAS Y TECNOLOGIAS Y TE VISIONAN TELEVISIONES DE PLASMA GIGANTES Y TE SUGUENN PASO A PASO CON LO K HACES Y SI ESTAS HACIENDO COSAS EN INTERNET SOBRE ELLOS...PORQUE SU PRINCIPAL TRABAJO SOMOS NOSOTROS...HH...LOS QUE PRETENDEN VIVIR UNA VIDA NORMAL CONSENSUAL SON EN REALIDAD SONAMBULOS Y ZOMBIES...A MERCED DE LAS ONDAS DE TRACCION DEL REPTILUANO MAYOR EN UNA DE ESAS CIUDADES SUB ESTALAGEM O SUS SECUACES Y AYUDANTES EN LA SUPERFICIE...HH.

#HOY HE SOÑADO CON EVA...IBAMOS A UNA ESCUELA EN RUZAFA Y TODO EL TIEMPO LA COGIA DE LA CINTURA...EL MEJOR SUEÑO DE MI VIDA..HUAHYAHUA!!! 2 09 2015 CAL GREG...HH
HACER UN VIDEO DEL CACTUS REPTANTE DE PATRICIA Y COLOCARLO EN VIMEO O YOUTUBE O OTRA PLATAFORMA PERO K SE VEA O UNAS FOTOS...HH IMPRESIONARA EN EL FB...HH

e - Commander X).

"Our explorer J.D. (name on file - Commander X), who is a mountain guide of the Mystery Mountain near Joinville (where there is supposed to be an entrance) said that, several times, he saw a luminous flying saucer ascend from the tunnel opening that leads to a subterranean city inside the mountain, in which he heard the beautiful choral singing of men and women, and also heard the 'canto galo' (rooster crowing), a universal symbol indicating the existence of subterranean cities in Brazil. He said that the saucer was so luminous that it lit up the night sky and converted it into daylight. On one occasion he met a group of subterranean men outside the tunnel. They were short, stocky, with reddish beards and long hair, and very muscular. When he tried to approach them, they vanished. Often he saw strange illuminations in this area at night which were probably produced by flying saucers (We use the name 'Mystery Mountain,' rather than reveal the true name of the mountain, so that unwanted outsiders will not come here to locate it). Throughout my many years of research I have accumulated a vast amount of data which would indicate that these entrances to subterranean cities abound throughout the region.

"An elderly man living in Joinville once told me that he had visited a tunnel near Concepiao in the state of Sao Paulo, and saw in the distance a marvelous subterranean city with vehicles darting back and forth, evidently traveling through tunnels from one subterranean city to another.

"Although the following report requires confirmation, it was told to me by an explorer named N.C. who said that he had visited a tunnel near Rio Casdor and had met a beautiful young woman appearing to be about 20 years of age. She spoke to him in Portuguese and SAID that she was 2,500 years old. He also met a bearded subterranean man

(Note: Often humans encountered in aerial disks or subterranean caverns declare that they are extremely old by humans standards. On the surface this might sound next to impossible, unless a revolutionary scientific breakthrough on the part of these human 'aliens' has allowed them to retard the ageing process to an extreme degree, or could the possibly that they are separated from the degenerating radiations of solar rays explain their allegedly greater longevity? Another possibility would be that throughbionics/biological transplants/prosthetics, etc. the lifespan of human beings possessing advanced biological and technological sciences might theoretically be increased dramatically. Incidentally, the writer and traveller Robert Stacy-Judd in some of his booksdescribed an exploration he and others in his party made of the peripheral areas of the Loltun caves of Yucatan.

Legend says that at least one group of people, fleeing persecution, entered en masse into the massive Loltun caves and were never seen again. Stacy-Judd tells of his own encounter with a 'cave hermit' deep in the cavern chambers who claimed to be well over 1000 years old, and who said he was a guardian of the cave and of the treasures--and city?--which lay deep below in the unknown depths, 'unknown' that is, except to the strange 'hermit'. Aside from photographs of this hermit which appeared in some of his works, the author also revealed photographs of 'underground gardens' consisting of areas of the cave which contain small patches of 'jungle', watered and lit through parts of the cavern ceilings which had collapsed, exposing them to the outer world. Whether such claims of longevity are real or whether the "subterranean" people were just playing with the minds of such explorers who encountered them, is uncertain - Branton).

#SUNSHINE HAVE BLOWN MY MIND AND THE WIND BLOWING MY BRAIN...HH.
#ESTOY EN EL PAIS DE LOS ZOMBIES VAMPIROS...HUAHUAHUA!!!
#M.I.A.B.23:00MECHH.TMR.G.O.#
THE SALVATION OF THE UNIVERSE...THE RETURN OF AL ANDALUS (A.A.)...HH.
#SUNSHINE HAVE BLOWN MY MIND AND THE WIND BLOWING MY BRAIN...HH.
#ESTOY EN EL PAIS DE LOS ZOMBIES VAMPIROS...HUAHUAHUA!!!

#M.I.A.B.23:00MECHH.TMR.G.O.#
THE SALVATION OF THE UNIVERSE...THE RETURN OF AL ANDALUS (A.A.)...HH.
#SUNSHINE HAVE BLOWN MY MIND AND THE WIND BLOWING MY BRAIN...HH.
#EN LO QUE OCURRIO EN EL DESIERTO DE OUARZAZAT LAS RELIGIONES NO TIENEN NADA QUE VER...NINGUNA...ESTO QUE ES LA REALIDAD ESTÁ MAS ALLA..DE LAS RELIGIONES DE TODAS...CLARO!!!...HEHE!!!.
Y ES LA REALIDAD...ES ALGO QUE SON EMBARGO ESTA TOTALMENTE ENRAIZADO EN NUESTRA ALMA EN LO MAS PROFUNDO Y AHORA ES EL TIEMPO DE SACAR A LA LUZ ESA HISTORIA ESA LUCHA...Y TU ERES PROTAGONISTA SOPHIA TU ERES LA HIJA DEL DESIERTO...NO LO OLVIDES...HH.
#HABLAR EN "NEXUS" DE LO DEL PRIMER MINISTRO GRIEGO.
Y SI CLON...INCLUIRÑO Y LO DE EVA Y LOS EMBARAZOS DE 3 MESES....HH
Alain Villeneuve...hh.
#HABLAR EN "NEXUS" DE LO DEL PRIMER MINISTRO GRIEGO.
Y SI CLON...INCLUIRÑO Y LO DE EVA Y LOS EMBARAZOS DE 3 MESES....HH.
PONER UN VÍDEO DE YOUTUBE DE LA VALENCIA DE RICARDO PEREZ CASADO.
jEXPERIENCIAS GENETICAS REPTILIANAS...NEXUS...FOTO...#geneticalreptiliansxxxc en LENOVO.Hum
BAJAR METABARONES Y VER LOS COMICS EN EL JATDIN PARA REDACTAR "NEXUS" MUY EN EL CONTEXTO DE LA VALENCIA DE 1984 CONTEXTUALUZ EL LIBRO AHI Y EN LA VALENCIA D PEREZ CASADO PERO SIN PERDERTE EN LOS DETALLES HISTORICOS SINO EN LA PSICOLOGIA DE LOS PERDONSJES MAS COMO UNA EXCUSA QUE COMO UNA INTROD. HISTORICA Y HAWAIKA...COMICS METAL HURLANT...Y ESCRITO EN BRASIL...COMO AHORA EXORZIZAR LOS MOV REPTILISNOS DE AHORS MISMAS ARMAS METAPSICOLOGICAS...AH Y LO MAS IMPORRANTE RECORDAR QUE ESTAMOS EN UN AÑO ABEJA-ARTE (GAY) Y QUE ES UN AÑO NO PARA CAMBIAR NADA SINO PARA CREAR..Y LEER OTRA VEZ "EL GRAN PLAN DIVINO" OBLIGAO!!!... 😊😊😊😊 .SIN LIMITACIONES...HH..PLAZA DE OTROS ESTACION DEL NORTE MALVAROSA DISEÑADOR DE VALENCIA MONTESINOS DISCOTECA DE LA MALVAROSA...NOMBRE...HH PERO VOLVER AQUI PARA VOLVER OTRA VEZ ALLI Y CAMBIARLO TODO LA HISTORIA...HH Y DE NUEVO VUELTA AHORA A LA GALAXIA LYRA CON XENUS....Y BRASIL 2015 EN RIO GRANDE DO SUL ESA TEXTURA...HH...METER PROTEUS EN VALENCIA...HH...EL AÑO 1981 FUE OTRO AÑO DE METAL...HH EL AÑO K VIENE SI SERA UN AÑO DE VIAJES MUCHOS VIAJES...HH.
#HAY MUCHAS RAZAS GRISES...UNAS SON AMISTOSAS Y OTRAS SON REPTILIANAS...HH
COLOCAR AQUI "TODAS" LAS FOTOS DEL PENDRIVE QUE TRAJO MI MADRE PORQUE FORMAN PARTE DE LA HISTORIA..LUEGO HAREMOS LA SELECCION...HEHE!!!.

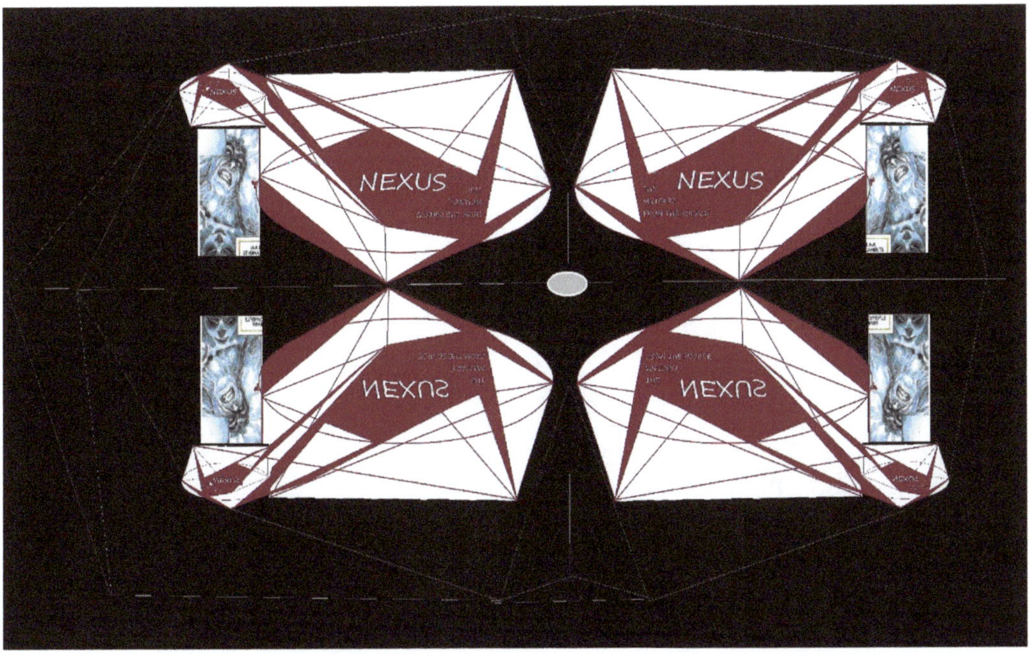

PASAR NOTAS AHK :

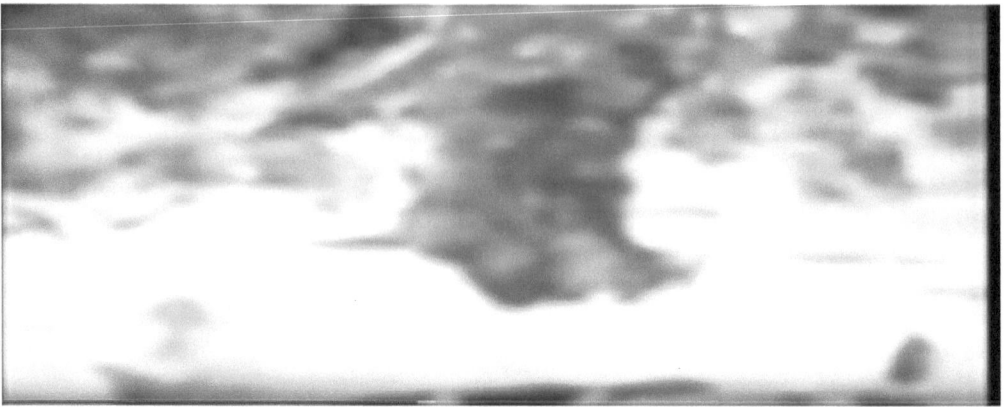

Rocks from steven the nephilin against fallen angels.
ZONA 84 NUMNERO 61 PAG 61..EL LIBRO COBRA UN NUEVO E INESPERADO GIRO…NOS VEMOS ABOCADOS A LA ESPAÑA DE LA GUERRA CIVIL 1936-37-38-39…HEHE!! ALLI SERÁ DONDE TENDRÁ LUGAR LA ACCION LOS ESCENARIOS LOS PERSONAJES…PERFECTO..
LOS ERIDANIOS EN LA GUERRA CIVIL…QUE BUENA IDEA!!!...DISFRAZADOS DE REPUBLICANOS O LAS BRIGADAS INTERNACIONALES LUCHARAN CONTRA LAS TROPAS DE FRANCO HASTA LA MUERTE…HH..BUENO ESO YA SE INTENTO CON "EL LABERINTO DEL FAUNO"..SI!!!...PERO ESTA VEZ ES MAS COMPLETO SIENDO TODAS LAS HISTORIAS UNA SOLA HISTORIA…SOBRE EL CONTEXTO DE LA GUERRA CIVIL…Y UN FINAL SORPRENDENTE….PODER RGERESAR A ESA EPOCA Y CAMBIAR EL RESULTADO DE LA CONTIENDA…HH….Y EL NACIMIENTO DE NEXUS XENUS EN ESPAÑA A LO PROTEUS…A TRAVESDE UNA MAQUINA PENSANTE ESCONDIDA EN UN CASTILLO O CUEVA O SUBTERRANEO DE MADRID…HH
1700 ANO DOMINI ROGUE RACE THE SHADOW PEOPLE 1936 SE LES VIO EN LAS COSTAS DE ALMERIA FRAGA SE BAÑO EN ESAS MISMAS AGUAS AÑOS DESPUES POR LOS MISMOS MOTIVOS
 EXOCOSMOBIOLOGIA EN LA GUERRA CIVIL ESPAÑOLA…ROGUE VLASH MAITRE..RPETILIANS GRISES…HEHE!!!
EN LAS ESTRIBACIONES DE LA SIERRA DE GUADARRAMA…HEHE!!!

Cyrax y Flyrax son clones de los Grises
Regane,prisión subterránea en el desierto de Argelia donde los rep-grises realizan experimentos con humanos,y esta temática está relacionada con la "Virtual Reality"..a solucionar Twitter-Hotmail.
Filme "Virus",èbola HV1 Grupo Austríaco llamado "Sabotaje" que reciclaron armas sónicas de los ejéritos soviets para crear máquinas de sonido absolutamente increibles..
La Tecnocracia ha sido la EXCRECENCIA o la manifestaión mal entendida de la izquierda, la energia flotante de la estructura del Soialismo Real que mantenía al sistema del NWO controlado..
Subir a MEGA todos los videos de "El Pícaro" ..
Dr Jiménez del oso y los "personajes" del Youtube.
Radio Tanambi..programas científicos[Buckminster Füller]
Porque "PODEMOS" no habla de Juan Carlos Torraija.
Instalar el Audio Pocket.
El jugo de una araña carangujeira brasileña encerrad en un bote después de 7 dias da "Omnisciencia",y gran apetito sexual…hehe!!!.

LISTADO DE LINKS/BIBLIOGRAFIA :

Todas las texturas de mis programas de radio …Estos son los Links de los programas grabados de "# radio tanambi on air" nuestro programa de radio en internet,nos podéis encontrar todos los días en http://mixlr.com/aurelio-perez-devon :
POR ORDEN CRONOLOGICO :
1- http://mixlr.com/aurelio-perez-devon/aurelio-perez-devon-on-mixlr-34
2- http://mixlr.com/aurelio-perez-devon/aurelio-perez-devon-on-mixlr-35
3- http://mixlr.com/aurelio-perez-devon/radio-tanambi-on-airhehe
4- http://mixlr.com/aurelio-perez-devon/aurelio-perez-devon-on-mixlr-36
5- http://mixlr.com/aurelio-perez-devon/aurelio-perez-devon-on-mixlr-40
6- http://mixlr.com/aurelio-perez-devon/aurelio-perez-devon-on-mixlr-41
7- http://mixlr.com/aurelio-perez-devon/aurelio-perez-devon-on-mixlr-42
8- http://mixlr.com/aurelio-perez-devon/aurelio-perez-devon-on-mixlr-43
9- http://mixlr.com/aurelio-perez-devon/aurelio-perez-devon-on-mixlr-47
10- http://mixlr.com/aurelio-perez-devon/aurelio-perez-devon-on-mixlr-48
11- http://mixlr.com/aurelio-perez-devon/aurelio-perez-devon-on-mixlr-49
12- http://mixlr.com/aurelio-perez-devon/aurelio-perez-devon-on-mixlr-50
13- http://mixlr.com/aurelio-perez-devon/aurelio-perez-devon-on-mixlr-51
14- http://mixlr.com/aurelio-perez-devon/aurelio-perez-devon-on-mixlr-52
15- http://mixlr.com/aurelio-perez-devon/aurelio-perez-devon-on-mixlr-53
16- http://mixlr.com/aurelio-perez-devon/aurelio-perez-devon-on-mixlr-56
17- http://mixlr.com/aurelio-perez-devon/aurelio-perez-devon-on-mixlr-73
18- http://mixlr.com/aurelio-perez-devon/aurelio-perez-devon-on-mixlr-75
19- http://mixlr.com/aurelio-perez-devon/aurelio-perez-devon-on-mixlr-76
20- http://mixlr.com/aurelio-perez-devon/radio-tanambi-on-aircine-distopico-session
21- http://mixlr.com/aurelio-perez-devon/aurelio-perez-devon-on-mixlr-77
22- http://mixlr.com/aurelio-perez-devon/aurelio-perez-devon-on-mixlr-79
23- http://mixlr.com/aurelio-perez-devon/aurelio-perez-devon-on-mixlr-80
24- http://mixlr.com/aurelio-perez-devon/aurelio-perez-devon-on-mixlr-81
25- http://mixlr.com/aurelio-perez-devon/aurelio-perez-devon-on-mixlr-82
26- http://mixlr.com/aurelio-perez-devon/aurelio-perez-devon-on-mixlr-83
27- http://mixlr.com/aurelio-perez-devon/aurelio-perez-devon-on-mixlr-85
28- http://mixlr.com/aurelio-perez-devon/aurelio-perez-devon-on-mixlr-87
29- http://mixlr.com/aurelio-perez-devon/radio-tanambi-on-air-special-david-bowie
30- http://mixlr.com/aurelio-perez-devon/aurelio-perez-devon-on-mixlr-89
31- http://mixlr.com/aurelio-perez-devon/aurelio-perez-devon-on-mixlr-95
32- http://mixlr.com/aurelio-perez-devon/aurelio-perez-devon-on-mixlr-96
33- http://mixlr.com/aurelio-perez-devon/aurelio-perez-devon-on-mixlr-97
34- http://mixlr.com/aurelio-perez-devon/aurelio-perez-devon-on-mixlr-98
35- http://mixlr.com/aurelio-perez-devon/aurelio-perez-devon-on-mixlr-99
36- http://mixlr.com/aurelio-perez-devon/aurelio-perez-devon-on-mixlr-100
37- http://mixlr.com/aurelio-perez-devon/aurelio-perez-devon-on-mixlr-101
38- http://mixlr.com/aurelio-perez-devon/radio-tanambi-on-airspecial-christmasthehe
39- http://mixlr.com/aurelio-perez-devon/radio-tanambi-on-air-special-christmast
40- http://mixlr.com/aurelio-perez-devon/aurelio-perez-devon-on-mixlr-103
41- http://mixlr.com/aurelio-perez-devon/radio-tanambi-on-airspecial-motorhead
42- http://mixlr.com/aurelio-perez-devon/radio-tanambi-on-airspecial-highlander
43- http://mixlr.com/aurelio-perez-devon/radio-tanambi-on-air-special-nirvanahehe

REGALO!!! : LISTADO MIS LIBROS GRATIS EN PDF :

1- "LA EYZEHD-LYA,LA APERTURA DEL SELLO" :
https://play.google.com/books/reader?printsec=frontcover&output=reader&id=VsYMBwAAAEAJ&pg=GBS.PA0

DE VENTA EN :
https://www.createspace.com/4770310

2- " ARGÜELLES EN RECONSTRUCCIÓN,EL REINO DE LOS ÁNGELES" :
https://play.google.com/books/reader?printsec=frontcover&output=reader&id=nKkMBwAAAEAJ&pg=GBS.PA0

DE VENTA EN :
https://www.createspace.com/5096493

3- "LOS DEFENSORES DE EPSYLON ":
https://play.google.com/books/reader?printsec=frontcover&output=reader&id=4bgCBwAAAEAJ&pg=GBS.PA0

DE VENTA EN :
https://www.createspace.com/5176173

4- " LOS INTELECTUALES DEL SIGLO 1378" :
https://play.google.com/books/reader?printsec=frontcover&output=reader&id=GGoCBwAAAEAJ&pg=GBS.PA0

DE VENTA EN :
https://www.createspace.com/5302884

5- " EL MANUSCRITO DE JACQUES DE MOLAY" :
https://play.google.com/books/reader?printsec=frontcover&output=reader&id=A7ICBwAAAEAJ&pg=GBS.PA0

DE VENTA EN :

https://www.createspace.com/4770261

6- " CODEX NEXUS,6.0,LA LIBERACIÓN " :
https://play.google.com/books/reader?printsec=frontcover&output=reader&id=WxUCBwAAAEAJ&pg=GBS.PA1

DE VENTA EN :
https://createspace.com/5801859

7- "EPSYLON" :
https://play.google.com/books/reader?printsec=frontcover&output=reader&id=ECkCBwAAAEAJ&pg=GBS.PA0

DE VENTA EN :

https://www.createspace.com/4879981

8-" EL EVANGELIO DE CHUCK" :
https://play.google.com/books/reader?printsec=frontcover&output=reader&id=62UCBwAAAEAJ&pg=GBS.PA0

DE VENTA EN :

https://www.createspace.com/4779772

9-" EL LIBRO DEL OCTÓGONO" :

https://play.google.com/books/reader?printsec=frontcover&output=reader&id=zJICBwAAAEAJ&pg=GBS.PA0
 DE VENTA EN :
https://www.createspace.com/4862381

 10- "EL EJÉRCITO DE LOS GRISES " :
https://play.google.com/books/reader?printsec=frontcover&output=reader&id=dqcCBwAAAEAJ&pg=GBS.PA0

DE VENTA EN :

https://www.createspace.com/4796456

11- "AMANECER INSEKTO´69" :
https://play.google.com/books/reader?printsec=frontcover&output=reader&id=ZFkCBwAAAEAJ&pg=GBS.PA9
 DE VENTA EN :
https://www.createspace.com/4769771

12- "LA TIKRAZIA INSEKTO,LA GEOLOGIA NEURAL Y EL SIGLO XXIX" :
https://play.google.com/books/reader?printsec=frontcover&output=reader&id=SVwCBwAAAEAJ&pg=GBS.PA0
DE VENTA EN :
https://www.createspace.com/4769637

13-" PLANETA VACIO" :
https://play.google.com/books/reader?printsec=frontcover&output=reader&id=q-4BBwAAAEAJ&pg=GBS.PA0
DE VENTA EN :
https://www.createspace.com/5278096

14-" EL LIBRO DE LA GUERRA" ;
https://play.google.com/books/reader?printsec=frontcover&output=reader&id=l1cCBwAAAEAJ&pg=GBS.PA0
DE VENTA EN :
https://www.createspace.com/4809602

15-"LOS NIÑOS PERDIDOS" :
https://play.google.com/books/reader?printsec=frontcover&output=reader&id=VvIBBwAAAEAJ&pg=GBS.PA16
 DE VENTA EN :
 https://www.createspace.com/5036277

16-" LAS ALAS DE LA LIBELULA-PRESCIENCIA INSEKTO" :
https://play.google.com/books/reader?printsec=frontcover&output=reader&id=nu8BBwAAAEAJ&pg=GBS.PA0
DE VENTA EN :
https://www.createspace.com/4769602

17- "EXOCOSMOBIOLOGIA" ;
https://play.google.com/books/reader?printsec=frontcover&output=reader&id=xwYCBwAAAEAJ&pg=GBS.PA0

DE VENTA EN :
https://www.createspace.com/5229782

18-" NICOLAS TESLA,MENTE ALIEN" (ENGLISH) :
https://play.google.com/books/reader?printsec=frontcover&output=reader&id=F-gBBwAAAEAJ&pg=GBS.PA0

DE VENTA EN :
https://www.createspace.com/5305370

19- "LOS 13 DIAS QUE CAMBIARON EL MUNDO" :
https://play.google.com/books/reader?printsec=frontcover&output=reader&id=WT0QBwAAAEAJ&pg=GBS.PA0

DE VENTA EN :
https://www.createspace.com/4962697

20-"LA COLMENA DE LA REINA" :
https://play.google.com/books/reader?printsec=frontcover&output=reader&id=7ZUQBwAAAEAJ&pg=GBS.PA0

DE VENTA EN :

https://www.createspace.com/4773857